AF386476

Representations of Infinite-Dimensional and Braid Groups

Ergodicity, Irreducibility, and Linearity

Representations of Infinite-Dimensional and Braid Groups

Ergodicity, Irreducibility, and Linearity

Alexander Kosyak

National Academy of Sciences of Ukraine, Ukraine &
London Institute for Mathematical Sciences, UK

World Scientific

NEW JERSEY · LONDON · SINGAPORE · BEIJING · SHANGHAI · HONG KONG · TAIPEI · CHENNAI · TOKYO

Published by

World Scientific Publishing Europe Ltd.

57 Shelton Street, Covent Garden, London WC2H 9HE

Head office: 5 Toh Tuck Link, Singapore 596224

USA office: 27 Warren Street, Suite 401-402, Hackensack, NJ 07601

Library of Congress Cataloging-in-Publication Data
Names: Kosyak, Alexander author
Title: Representations of infinite-dimensional and braid groups : ergodicity, irreducibility, and
  linearity / Alexander Kosyak, National Academy of Sciences of Ukraine, Ukraine &
  London Institute for Mathematical Sciences, UK.
Description: New Jersey : World Scientific, [2026] | Includes bibliographical references and index.
Identifiers: LCCN 2026001690 | ISBN 9781800619258 hardcover |
  ISBN 9781800619265 ebook | ISBN 9781800619272 ebook other
Subjects: LCSH: Infinite-dimensional manifolds | Braid theory
Classification: LCC QA613.2 .K679 2026
LC record available at https://lccn.loc.gov/2026001690

British Library Cataloguing-in-Publication Data
A catalogue record for this book is available from the British Library.

For any available supplementary material, please visit
https://www.worldscientific.com/worldscibooks/10.1142/Q0566#t=suppl

Desk Editors: Eshak Nabi Akbar Ali/Srinidhi Murugan

Typeset by Stallion Press
Email: enquiries@stallionpress.com

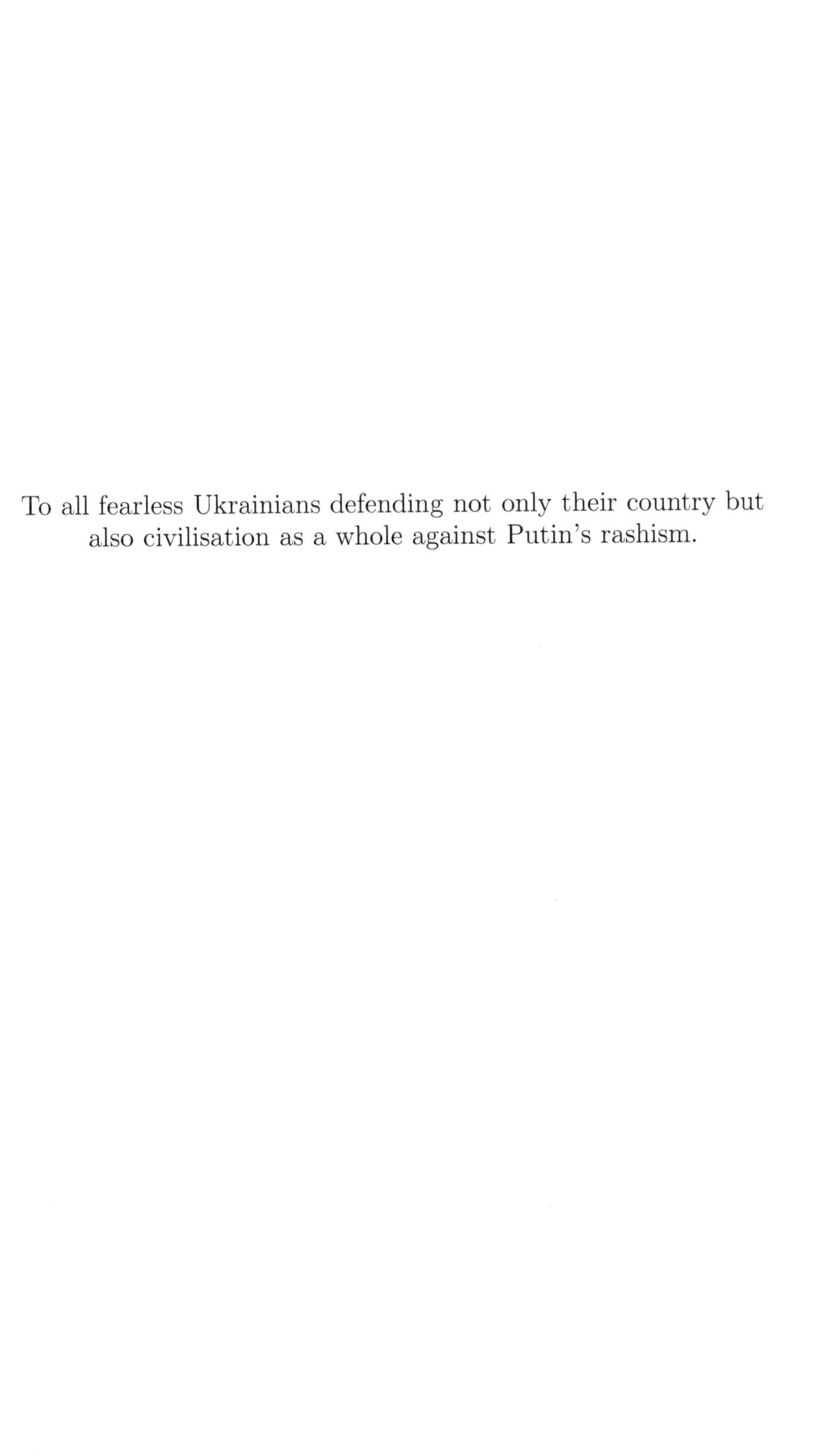

To all fearless Ukrainians defending not only their country but also civilisation as a whole against Putin's rashism.

Preface

In the *first part* of the book, we consider representations of infinite-dimensional non-locally compact groups. The *second part* of the book is devoted to representations of the braid groups B_n.

If a group is infinite-dimensional and non-locally compact, it does not admit a Haar measure. In a previous book by the author, titled *Regular, Quasi-Regular and Induced Representations of Infinite-Dimensional Groups*, an analogue of the regular, quasi-regular and induced representations was constructed and their irreducibility was studied. This book is written in the same spirit and explains how to prove irreducibility by studying the von Neumann algebra $\mathfrak{A}$ generated by the representations. We provide new insights connecting von Neumann algebras, ergodicity and irreducibility.

The content of the first part of the book is as follows. Chapter 1 contains a very brief introduction to the representation theory of locally compact groups. In Chapter 2, we explain our approach to representations of infinite-dimensional groups and give a brief history of the representations of the group $G = \varinjlim_n G_n$. In Chapter 3, we consider the group $G = \mathrm{GL}_0(2\infty, \mathbb{R})$, defined as the inductive limit of the general linear groups acting from the right on the space X of three infinite rows with the Gaussian product measure μ. The action involved "respects" the measure, i.e., μ^{R_t} is equivalent to μ for all $t \in G$, where R_t is the right action of G on X. The unitary representation of the group G on the space $L^2(X, \mu)$ appears naturally. We give a criterion for its irreducibility in terms of the action of the group $\mathrm{GL}(3, \mathbb{R})$ from the left on X in Chapter 4. Namely, we show that the representation is irreducible if and only if μ^{L_s} is orthogonal to μ for all $s \in \mathrm{GL}(3, \mathbb{R}) \setminus \{e\}$, where L_s is the left action of G on X. This is also a manifestation of a phenomenon predicted by the *Ismagilov conjecture*. To prove irreducibility (Chapter 4), we show that the von Neumann algebra generated by the representation contains certain abelian subalgebras. This inclusion follows from the orthogonality and can be seen as a kind of ergodic theorem (comparable to the law of large numbers but more subtle). More precisely, the elements of the corresponding commutative subalgebras can be approximated (in the strong resolvent sense) by the combinations of generators of one-parameter groups. This is described in Chapter 5. This approximation, being optimal at every finite step, represents the best possible outcome under the given conditions. Several facts about some estimates, the properties of an infinite parallelotope and the properties of the measures are collected in Chapter 6. For the approximation, we use properties of the generalised characteristic polynomial in Chapter 7. We also provide an explicit expression for the minimum of the quadratic form restricted to a hyperplane and a theorem regarding the height of an infinite parallelotope in Chapter 8. A possible generalisation of the results obtained for $m > 3$ is considered in Chapter 9. A few other interesting examples are considered in Chapter 10.

Our approach to representation theory is completely different from all previous approaches and is based on a variety of nonequivalent G-quasi-invariant measures on a G-space, (X, μ). To study irreducibility, we use *Ismagilov's conjecture*, which is a generalisation of *Dixmier's commutation theorem*. This approach allows us to prove that nonequivalent measures correspond to nonequivalent representations. Thus, the *nonequivalent measures become essential ingredients in the description of the dual* $\widehat{G}$ for infinite-dimensional groups G.

The second part of the book is devoted to representations of the braid groups B_n. The problem of *linearity* of the braid groups B_n was solved by Bigelow and Krammer; they used two-parameter representations, constructed by Lawrence.

Our aim is to explain how this representation is connected with the reduced Burau representation. We show that the *Lawrence–Krammer representation is the quantisation of the symmetric square of the reduced Burau representation*. The hint is the following: both representations of B_n have the same dimension $\frac{n(n-1)}{2}$, and their spectra coincide; see Remark 13.3. This connection allows us to construct new representations of the braid groups.

In 1936, Burau introduced the family of representations of the braid group B_n; see (12.1). The Burau representation for $n = 2, 3$ has been known to be *faithful*, i.e., *the kernel of the corresponding representation is trivial*.

The non-faithfulness for $n = 5$ was shown by Bigelow in 1999. The faithfulness of the Burau representation for $n = 4$ remains an *open problem*.

The Burau representation appears as a summand of the *Jones representation* (Jones, 1987), and for $n = 4$, the faithfulness of the Burau representation is equivalent to that of the Jones representation, which furthermore is related to the question of whether or not the Jones polynomial is an *unknot detector* (Bigelow, 2002b).

The problem of *linearity*, i.e., the existence of a faithful representation, for the braid groups B_n for $n \geq 4$ remained open until 2000. In 1990, Lawrence constructed a two-parameter family of representations of the braid group B_n, using homological methods. In 2000, Krammer, using completely algebraic methods, constructed the same representations and showed that these representations are faithful for B_4. One year later, in 2001, Bigelow showed that the Lawrence–Krammer representation is faithful for all B_n. The following year, Krammer (2002) proved the same result using a different method. Finally, it was established that the braid groups B_n are linear for all n: in particular, for $n \leq 3$, one can use the Burau representation, and for $n \geq 4$, one can use the Lawrence–Krammer representation.

We give the references separately for the first and second parts of the book at the end of each part.

A. V. Kosyak
London, Great Britain, 2026

About the Author

 **Alexander Kosyak** was born in the village of Pylypcha, Kyiv region, Ukraine. He studied mathematics at Taras Shevchenko National University of Kyiv (1972–1977). He then joined the Ukrainian Agricultural Academy as an assistant professor (1977–1982). He pursued postgraduate studies at the Institute of Mathematics of the National Academy of Sciences of Ukraine (NASU), Kyiv (1982–1985). After completing his Candidate of Sciences degree in 1985, he continued at the Institute of Mathematics as a Scientific Researcher (1985–2010). In 2010, he obtained his Doctor of Physics and Mathematical Sciences (habilitation) from the same institute, and in 2011 he was appointed Leading Scientific Researcher. Since 2023, he has been an Arnold Fellow at the London Institute for Mathematical Sciences, Royal Institution, London. He has also held research positions at Aix-Marseille Université, Université Claude Bernard Lyon 1 in France, and the Max Planck Institute in Bonn, Germany (2006–2022).

Acknowledgements

A. Kosyak is very grateful to Prof. K.-H. Neeb, Prof. M. Smirnov and Dr P. Moree for their personal efforts to make academic stays possible at their respective institutes. Kosyak visited the Max Planck Institute for Mathematics (MPIM) from March to April 2022 and from January to April 2023, the University of Augsburg from June to July 2022, and the University of Erlangen–Nuremberg from August to December 2022, all during the Russian invasion of Ukraine. Also, Prof. R. Kashaev kindly invited Kosyak to Geneva. Furthermore, the author would like to pay his respects to Prof. P. Teichner at MPIM for his immediate efforts to help mathematicians in Ukraine following the Russian invasion.

Since the spring of 2023, Kosyak has been an Arnold Fellow at the London Institute for Mathematical Sciences (LIMS), and he would like to express his gratitude to Mrs Myers Cornaby, Ms Ker Mercer, and, especially, Dr T. Fink (Director of LIMS) and Prof. Y.-H. He.

The author would also like to acknowledge the invaluable contributions made by Dr Pieter Moree of MPIM, Bonn.

Contents

14. How to Construct New Representations of B_n — **217**

Part 1

Representations of Infinite-Dimensional Groups

Chapter 1

Representations of Locally Compact Groups

1.1 What Is Representation Theory About?

The main problem in the representation theory (RT) of a topological group G is to find its *unitary dual* $\widehat{G}$, i.e., *the set of all irreducible unitary representations* of the group G *up to the equivalence relation* and decompose any representation into a direct sum or direct integral of irreducible ones. Almost all constructions in RT for *locally compact groups,* such as *regular, quasi-regular and induced representations,* are based on the existence of the *Haar measure* on a group G. These constructions allow us to find the unitary dual $\widehat{G}$ for almost all locally compact groups G, except, e.g., the group $\mathrm{SO}(p,q)$.

1.2 Compact Groups and the Regular Representation

The existence of the Haar measure was proved by Haar (1933). Weil (1953) proved a converse, namely that a group with a quasi-invariant measure that acts faithfully on the L^2-space is locally compact with respect to the strong operator topology from $U(L^2(G))$. The *right ρ (resp., the left λ) regular* representation $\rho, \lambda : G \to U(H)$ of the group

G is defined on the Hilbert space $H = L^2(G, h)$ by

$$(\rho_t f)(x) = f(xt), \ (\lambda_s f)(x) = (dh(s^{-1}x)/dh(x))^{1/2} f(s^{-1}x),$$

$$t, s, x \in G, \ f \in H. \tag{1.1}$$

Since $[\rho_t, \lambda_s] = 0$ for all $t, s \in G$, both representations are reducible. When G is compact, the decomposition of the right regular representation contains *all* the irreducible representations:

$$\rho = \bigoplus_{\kappa \in \widehat{G}} c_\kappa \rho_\kappa. \tag{1.2}$$

1.3 The Dixmier Commutation Theorem

Lemma 1.1 (Dixmier, 1969). *Let* $\mathfrak{A}_G^\rho = (\rho_t | t \in G)''$ *and* $\mathfrak{A}_G^\lambda = (\lambda_s | s \in G)''$ *be the* von Neumann algebras *generated by the right and left regular representations of a locally compact group G. Then,*

$$\left(\mathfrak{A}_G^\rho\right)' = \mathfrak{A}_G^\lambda. \tag{1.3}$$

This lemma is a *cornerstone* of our study of representations of infinite-dimensional groups since the *Ismagilov conjecture*, as described in what follows, is a far-reaching generalisation of this statement.

1.4 Koopman's Representation

In order to construct a unitary representation of a topological group G (locally compact or infinite-dimensional), we use some *G-space* (X, μ) with a "good" measure μ. To be more precise, let $\alpha : G \to \mathrm{Aut}(X)$ be a measurable action of a group G on a measurable space (X, μ) with G-quasi-invariant measure μ, i.e., $\mu^{\alpha_t} \sim \mu$ (where $\sim$ means equivalent) for all $t \in G$, where $\mathrm{Aut}(X)$ is the group of all measurable automorphisms of X. We use the notation $\mu^f(\Delta) = \mu\big(f^{-1}(\Delta)\big)$ for $f : X \to X$, where Δ is some measurable set in X. With these data, we associate the representation $\pi^{\alpha,\mu,X} : G \to U\big(L^2(X,\mu)\big)$ of the group G given by the formula

$$(\pi_t^{\alpha,\mu,X} f)(x) = \big(d\mu(\alpha_{t^{-1}}(x))/d\mu(x)\big)^{1/2} f\big(\alpha_{t^{-1}}(x)\big), \quad f \in L^2(X,\mu). \tag{1.4}$$

In the case of an invariant measure, this representation is called *Koopman's representation*. We keep the same name for representation (1.4).

1.5 Quasi-Regular Representations

When the group is locally compact but not compact, the regular representation is not sufficient to find all irreducible representations. One should generalise the regular representation; for example, consider a quasi-regular or an induced representation, as for the group $SL(2,\mathbb{R})$, see Lang (1975). A quasi-regular representation of the group G is a particular case of Koopman's representation (1.4), with $X = H \setminus G$ being the set of *right cosets* (or $X = G/H$, the set of *left cosets*), where H is some closed subgroup of G and μ is a G-quasi-invariant measure on X. In the case of $X = H \setminus G$, group G acts on X from the right, while in the case of $X = G/H$, it acts from the left.

1.6 Induced Representations

To construct an *induced representation* of the group G, we should fix a closed subgroup H of the group G and a unitary representation $S : H \to U(V)$ of the group H. The induced representation $\operatorname{Ind}_H^G S$ of group G is defined on the space $L^2(X, V, \mu)$, where $X = H \setminus G$; for details, see Mackey (1976).

Chapter 2

Representations of Infinite-Dimensional Groups

2.1 A Brief History of the Representations of Infinite-Dimensional Groups

The representation theory of infinite-dimensional groups is a broad area. We mention here only some results connected with unitary representations of inductive limits of classical groups and an interesting connection with *random matrices*. Using his *orbit method* developed in 1962, Kirillov described in 1973 all unitary irreducible representations of the group $U_\infty(H)$, the *completion in the strong operator topology of the group* $U(\infty) = \varinjlim_n U(n)$. The group $U_\infty(H)$ consists of all unitary operators of the form $1 + a$, where a is compact.

This approach was generalised by Ol'shanskiĭ for the inductive limits of other classical groups, $K(\infty) = \varinjlim_n K(n)$, where K is U, O or Sp. In Ol'shanskiĭ (1991), the complete classification of the so-called "tame" representations of the group $K(\infty)$ was obtained; see also Neeb (2013).

Nessonov (1986) proved that a previously known list of indecomposable spherical functions of the group $\mathrm{GL}(\infty)$ that are bilaterally invariant with respect to the unitary subgroup is complete. Bufetov (2014) showed that a Borel measure on the space of infinite Hermitian matrices that is invariant under the action of the infinite unitary group under additional conditions must be finite.

7

The aim of the book by Stratila and Voiculescu (1975) is the profound study of the factor representations of the group $U(\infty)$.

Neeb (2017) describes the recent progress in the classification of bounded and semibounded representations of infinite-dimensional Lie groups. He begins with a discussion of the semiboundedness condition and explains how the new concept of a smoothing operator can be used to construct C^*-algebras (the so-called host algebras), whose representations are in one-to-one correspondence with certain semibounded representations of an infinite-dimensional Lie group G. This makes the full power of C^*-theory available in this context. Then, he discusses the classification of bounded representations of several types of unitary groups on Hilbert spaces and gauge groups. After explaining the method of holomorphic induction as a means to transition from bounded representations to semibounded ones, he describes the classification of semibounded representations for hermitian Lie groups of operators, loop groups (with infinite-dimensional targets), the Virasoro group and certain infinite-dimensional oscillator groups. Ol'shanskiĭ (1990) addresses the representation theory of the automorphism groups of infinite-dimensional Riemannian symmetric spaces. The book by Ismagilov (1996) is devoted to the representations of certain classes of infinite-dimensional Lie groups: current groups, diffeomorphism groups and some of their semidirect products.

Let $S_\infty = \cup_{n \geq 1} S_n$ be the group of finite permutations of natural numbers. All indecomposable central positive definite functions on S_∞, which are related to factor representations of II_1, were given by Thoma (1964). Later, Vershik and Kerov obtained the same result using a different method in Vershik and Kerov (1981) and provided a realisation of the representations of type II_1 in Vershik and Kerov (1981). In Kerov, Ol'shanskiĭ and Vershik (1993), the generalised regular representations $\{T_z : z \in \mathbb{C}\}$ of the group $S_\infty \times S_\infty$ were studied. These representations are deformations of the biregular representation of S_∞ in $l^2(S_\infty)$. A two-parameter family of the generalised regular representations $T_{z,z'}$ of the group S_∞ was also considered in Kerov, Ol'shanskiĭ and Vershik (1993).

In Borodin and Ol'shanskiĭ (1998), the corresponding spectral measure $P_{z,z'}$ was investigated. The correlation functions are of a determinantal form, similar to those studied in *random matrix theory.*

Borodin, Okounkov and Ol'shanskiĭ (2000) studied the asymptotics of the *Plancherel measures* M_n for the symmetric groups S_n. They showed that M_n converges to the delta measure supported on a certain subset Ω of $\mathbb{R}^2$ closely connected to *Wigner's semicircle law* for the distribution of eigenvalues of random matrices. In particular, they gave a positive answer to the conjecture proposed by Baik, Deift and Johansson (1999).

2.2 Our Approach to Representations of Infinite-Dimensional Groups

We consider infinite-dimensional non-locally compact groups. If the group is not locally compact, it does not admit a Haar measure; see Weil (1953). The main idea of Kosyak (2018) is to construct an *analogue* of the regular, quasi-regular and induced representations and study their irreducibility. Our approach to representation theory is completely different and is based on a *variety* of non-equivalent G-quasi-invariant *measures* μ on a G-space (X, μ). To study irreducibility, we use the Ismagilov conjecture, a generalisation of Dixmier's commutation theorem. This approach allows us to prove that *nonequivalent measures* correspond to *nonequivalent representations*; see Remark 2.1. Thus, the *nonequivalent measures* become *essential ingredients* in the description of the *dual* $\widehat{G}$ for infinite-dimensional groups G.

2.3 The Ismagilov Conjecture

In order to construct an *analogue of the regular representation* of an infinite-dimensional group G, we can first try to find a triplet

$$(\widetilde{G}, G, \mu), \tag{2.1}$$

where $\widetilde{G}$ is some larger topological group containing G as a dense subgroup, and a measure μ on $\widetilde{G}$ which is right or left G-quasi-invariant, i.e., $\mu^{R_t} \sim \mu$ for all $t \in G$, (or $\mu^{L_s} \sim \mu$ for all $s \in G$). Here, $\sim$ means *equivalence*; for details, see Kosyak (2018). Consider the right and left actions R_t, L_s of the group G on $\widetilde{G}$, defined as follows:

$$R_t x = x t^{-1}, \quad L_s x = sx, \quad t, s \in G, \ x \in \widetilde{G}.$$

Denote by μ^{R_t}, μ^{L_s} the images of the measure μ under the maps $R_t, L_s : \widetilde{G} \to \widetilde{G}$. The right and left representations $T^{R,\mu}, T^{L,\mu} : G \to U(L^2(\widetilde{G}, \mu))$ are naturally defined on the Hilbert space $L^2(\widetilde{G}, \mu)$ by the following formulas (compare with (1.1)):

$$(T_t^{R,\mu} f)(x) = (d\mu(xt)/d\mu(x))^{1/2} f(xt), \tag{2.2}$$

$$(T_s^{L,\mu} f)(x) = (d\mu(s^{-1}x)/d\mu(x))^{1/2} f(s^{-1}x). \tag{2.3}$$

In Kosyak (2018, Chapter 5, Theorem 5.2.11), we proved that for the group $B_0^{\mathbb{N}} = \varinjlim_n B(n, \mathbb{R})$, where $B(n, \mathbb{R})$ is the group of upper-triangular real matrices with units on the diagonal and a Gaussian product measure μ_b on the group $B^{\mathbb{N}}$, the Dixmier commutation theorem holds when $\mu_b^{L_s} \sim \mu_b$ for all $s \in B_0^{\mathbb{N}}$ under some special conditions on the measure μ_b. Here, $B_0^{\mathbb{N}}$ (resp., $B^{\mathbb{N}}$) is a group of infinite real matrices of the form $I + x$, where x is upper-triangular with a finite number of nonzero elements (resp., $x = \sum_{k<n} x_{kn} E_{kn}$ is arbitrary upper-triangular) and

$$\mu_b(x) = \bigotimes_{k<n} \mu_{(b_{kn},0)}(x_{kn}), \quad d\mu_{(b_{kn},a_{kn})}(x_{kn})$$

$$= \sqrt{\frac{b_{kn}}{\pi}} e^{-b_{kn}(x_{kn}-a_{kn})^2} dx_{kn}. \tag{2.4}$$

However, the right regular representation of an infinite-dimensional group can be irreducible if no left actions are *admissible* for the measure μ, i.e., when $\mu^{L_s} \perp \mu$ for all $s \in G \backslash \{e\}$. In this case, the von Neumann algebra $\mathfrak{A}^{T^{L,\mu}}$ generated by the left regular representation $T^{L,\mu}$ is trivial.

Conjecture 1 (Ismagilov, 1985). *The right regular representation defined by (2.2)*

$$T^{R,\mu} : G \to U(L^2(\widetilde{G}, \mu))$$

is irreducible if and only if:

(1) $\mu^{L_s} \perp \mu$ *for all* $s \in G \backslash \{e\}$, *(where* $\perp$ *stands for singular or orthogonal measures);*

(2) *the measure* μ *is* G*-ergodic.*

Recall that the probability measure μ on some G-space X is called *ergodic* if any function $f \in L^1(X, \mu)$, with property $f(\alpha_t(x)) = f(x) \bmod \mu$ for all $t \in G$, is constant. Here, $\alpha : G \to \operatorname{Aut}(X)$ is the action of the group G on the space X; see details in the following section. Conditions (1) and (2) are the necessary irreducibility conditions, at least for Gaussian measures. The challenge is to prove that they are sufficient too.

Remark 2.1. Conjecture 1 was expressed by Ismagilov in his referee report of the author's PhD thesis in 1985. It was verified for a large number of particular cases. In Kosyak (1992) (see also Kosyak, 2018, Theorem 2.1.1), we proved Conjecture 1 for the group $B_0^{\mathbb{N}}$ and a Gaussian product measure μ_b on the group $B^{\mathbb{N}}$. Moreover, we proved that two irreducible representations T^{R,μ_b} and $T^{R,\mu_{b'}}$ are equivalent if and only if the corresponding measures μ_b and $\mu_{b'}$ are equivalent (Kosyak, 2018, Theorem 2.1.17). In the general case, Conjecture 1 remains an open problem; for details, see Kosyak (2018).

2.4   Irreducibility of Koopman's Representation

If a G-space X has a "natural" right action of the group G and a left action of another group G_1, such as in the case of the Schur–Weil duality, and these actions commute, $[R_t, L_s] = 0$ for all $t \in G, s \in G_1$, then we can imagine that the Koopman representation (1.4) is irreducible if the left action is not admissible, i.e., $\mu^{L_s} \perp \mu$ for all $s \in G_1 \setminus \{e\}$, and the measure μ is G-right-ergodic, such as in the Ismagilov conjecture. The main result of the first part of the book, Theorem 3.1, is a particular case of this situation. However, if we have only one action $\alpha : G \to$ $\mathrm{Aut}(X)$, the right action $R(G)$ should be replaced by $\alpha(G) \subset \mathrm{Aut}(X)$ and the left action $L(G_1)$ by the *centraliser* of the subgroup $\alpha(G)$ in the group $\mathrm{Aut}(X)$. The following conjecture is a natural generalisation of the Ismagilov conjecture.

Conjecture 2. *The representation* (1.4) *is irreducible if and only if:*

(1) $\mu^g \perp \mu$ *for all* $g \in Z_{\mathrm{Aut}(X)}(\alpha(G)) \backslash \{e\}$;
(2)   *the measure* μ *is* G-*ergodic.*

Here, $Z_G(H)$ is the *centraliser* of the subgroup H in the group G: $Z_G(H) = \{g \in G \mid \{g, a\} = e$ for all $a \in H\}$, where $\{g, a\} = gag^{-1}a^{-1}$. In general, Conjecture 2 is false. In the case of a finite field $\mathbb{F}_p$, we need some additional conditions for irreducibility (Kosyak, 2016). Our aim is to determine them.

2.5   Representations of the Inductive Limit Groups $G = \lim_n G_n$

Consider an inductive limit $G = \varinjlim_n G_n$ with all $\widehat{G_n}$ known.

Problem 2.1. Is it sufficient to determine $\widehat{G}$, i.e., whether $\widehat{G} = \varprojlim_n \widehat{G_n}$?

In the case of commutative groups $G_n = \mathbb{R}^n$ or $G_n = T^n = T \times \cdots \times T$, the answer to Problem 2.1 is positive. In the first case, we have $\widehat{G} = \varprojlim_n \widehat{G_n} = \mathbb{R}^\infty$. In the second, we have $\widehat{G} = \varprojlim_n \widehat{G_n} = \mathbb{Z}^\infty$. If we have $\rho_n \in \widehat{G_n}$ defined on the Hilbert space H_n and there is an embedding of Hilbert spaces $i_n : H_n \to H_{n+1}$, we can define the representation $\rho = \varinjlim_n \rho_n$ in the space $H = \varinjlim_n H_n = \cup_{n \in \mathbb{N}} H_n$, and this representation should be irreducible. But usually, the space $H = \varinjlim_n H_n$ *has no Hilbert structure*; see, e.g., Wolf (2004), as on the space $\mathbb{R}_0^\infty = \varinjlim_n \mathbb{R}^n$. However, when the representations $T = \varinjlim_n T_n$

can be obtained as limit of Koopman's representations $T_n = \pi^{\alpha_n, \mu_n, X_n}$ on the space $H_n = L^2(X_n, \mu_n)$ and we have an additional structure:

$$X_n = X^{(1)} \times \cdots \times X^{(n)}, \quad \mu_n = \bigotimes_{k=1}^{n} \mu^{(k)}, \qquad (2.5)$$

so that the final object $\varinjlim_n H_n$ can be embedded in a Hilbert space $\otimes_{k=1}^{\infty} H^{(k)}$. The construction of the *von Neumann infinite tensor product* of the Hilbert spaces $H^{(k)}$,

$$\mathcal{H}_e = \bigotimes_{k=1,e}^{\infty} H^{(k)}, \qquad (2.6)$$

can be found in Berezanskii (1986) (see also Khrennikov, Kosyak and Shelkovich (2012)). Here, $e = (f_k)_{k=1}^{\infty}$ is some *stabilisation sequence*, where $f_k \in H^{(k)}$. Two infinite products, $\mathcal{H}_e$ and $\mathcal{H}_l$, corresponding to two different stabilisations are *equivalent* if and only if the *corresponding stabilisations are equivalent $e \sim l$*.

As was shown in Kosyak (2018, Chapter 8), the answer to Problem 2.1 is negative, at least for the group $B_0^{\mathbb{N}} = \varinjlim_n G_n$, where $G_n = B(n, \mathbb{R})$. The dual $\widehat{G_n}$ is described by the orbit method, but with this procedure, we cannot obtain all irreducible representations, in particular, the regular ones. According to Kosyak (2018, Theorem 2.1.1), the representation $T^{R, \mu_b} : B_0^{\mathbb{N}} \to U\big(L^2(B^{\mathbb{N}}, \mu_b)\big)$ is irreducible if and only if $\mu_b^{L_s} \perp \mu_b$ for all $s \in B_0^{\mathbb{N}} \setminus \{e\}$, where μ_b is defined by (2.4). Denote by R_n the regular representation of the group $B(n, \mathbb{R})$ in the Hilbert space $H_n = L^2(G_n, h_n)$, where h_n is the Haar measure on G_n. But the restriction $T^{R, \mu_b}|_{G_n}$ is equivalent to the regular representation R_n of G_n, all of which are reducible. We have $T^{R, \mu_b} = \varinjlim_n R_n$; for details, see Kosyak (2018, Chapter 2.4). *So, we can obtain the irreducible representations as inductive limits of reducible ones.* Moreover, we prove (Kosyak, 2018, Theorem 2.1.17) that two irreducible representations, T^{R, μ_b} and $T^{R, \mu_{b'}}$, are equivalent if and only if the corresponding measures μ_b and $\mu_{b'}$ are equivalent. Nonequivalent measures, in fact, give us two nonequivalent infinite tensor products, $\mathcal{H}_e \not\sim \mathcal{H}_l$. This means that different embeddings of the spaces $i_n : H_n \to H_{n+1}$ can give nonequivalent representations; see details in Kosyak (2018, Chapter 8).

2.6 The Inductive Limit of Reducible Koopman Representations Can Be Irreducible

In this section, we consider the Koopman representations of the inductive limit of the general linear group $\mathrm{GL}_0(2\infty, \mathbb{R}) = \varinjlim_{n,i^s} \mathrm{GL}(2n+1, \mathbb{R})$ with respect to the symmetric embedding (3.2). Theorem 3.1 states that $T^{R,\mu,3}$ is irreducible if and only if $(\mu_{(b,a)}^m)^{L_s} \perp \mu_{(b,a)}^m$ for all $s \in \mathrm{GL}(3, \mathbb{R})\backslash\{e\}$. But the restriction $T^{R,\mu,3}|_{G_n}$ of the representation $T^{R,\mu,3}$ to the subgroup G_n is the Koopman representation of G_n, which is reducible since any action of the group $\mathrm{GL}(3, \mathbb{R})$ from the left on the space $X_{3,n}$ is admissible, i.e., $\left(\mu_{(b,a)}^{3,n}\right)^{L_s} \sim \mu_{(b,a)}^{3,n}$ for all $s \in \mathrm{GL}(3, \mathbb{R})$.

2.7 The General Idea to Prove Irreducibility

Let G be some infinite-dimensional group acting on a G-space (X, μ) equipped with some quasi-invariant measure. In the concrete examples considered in Kosyak (1992)–Kosyak (2023), the possibility to approximate a lot of functions in $L^\infty(X, \mu)$ using Lemma 6.1, follows from the fact that

$$\lim_{n \to \infty} (C_n(\lambda)^{-1} a_n, a_n) = \infty, \quad \text{where} \quad C_n(\lambda) = \mathrm{diag}(\lambda_1, \ldots, \lambda_n) + C_n.$$

$$(2.7)$$

According to Theorem 5.3 proved in Kosyak (2023) (see also Lemma 6.5 for $m = 3$), we have

$$\left(C_n(\lambda)^{-1} a_n, a_n\right) = \Delta(y_1^{(n)}, y_2^{(n)}, \ldots, y_m^{(n)})$$

$$= \frac{\det\left(I_m + \gamma(y_1^{(n)}, y_2^{(n)}, \ldots, y_m^{(n)})\right)}{\det\left(I_{m-1} + \gamma(y_2^{(n)}, \ldots, y_m^{(n)})\right)} - 1. \quad (2.8)$$

Finally, by Lemma 1.2 (Kosyak, 2025a) (see Lemmas 6.4 and 6.6 for $m = 3$), we get

$$\lim_{n \to \infty} \frac{\det\left(I_m + \gamma(y_1^{(n)}, y_2^{(n)}, \ldots, y_m^{(n)})\right)}{\det\left(I_{m-1} + \gamma(y_2^{(n)}, \ldots, y_m^{(n)})\right)} = \infty.$$

The first part of the book is organised as follows. The main result and the idea of the proof are formulated in Section 3.2, and the orthogonality problem in measure theory is studied in Section 3.3. Irreducibility is considered in Chapter 4, and the approximation of x_{kn} or D_{kn} is considered in Chapter 5. Some useful facts are presented in Chapter 6. In Chapter 7, we explain the use of *generalised characteristic polynomials*. In Chapter 8, we give the explicit expression for the minimum of the quadratic form restricted to a hyperplane and present a result of independent interest on the *height of an infinite parallelotope*. In Chapter 9, we provide some hints on how to deal with the case of $m > 3$. In Chapter 10, we present a few more interesting examples.

Chapter 3

Representations of the Group $\mathrm{GL}_0(2\infty, \mathbb{R})$

3.1 Finite-Dimensional Case

Consider the space

$$X_{m,n} = \left\{ x = \sum_{1 \leq k \leq m} \sum_{-n \leq r \leq n} x_{kr} E_{kr}, \ x_{kr} \in \mathbb{R} \right\},$$

where E_{kn}, $k, n \in \mathbb{Z}$ are infinite matrix unities, with the measure (see (3.5))

$$\mu_{(b,a)}^{m,n}(x) = \bigotimes_{k=1}^{m} \bigotimes_{r=-n}^{n} \mu_{(b_{kr}, a_{kr})}(x_{kr}).$$

Two groups act on the space $X_{m,n}$, namely $\mathrm{GL}(m, \mathbb{R})$ from the left and $\mathrm{GL}(2n + 1, \mathbb{R})$ from the right, and their actions commute. Therefore, two von Neumann algebras $\mathfrak{A}_{1,n}$ and $\mathfrak{A}_{2,n}$ in the Hilbert space $L^2(X_{m,n}, \mu_{(b,a)}^{m,n})$ generated, respectively, by the left and right actions of the corresponding groups have the property that $\mathfrak{A}_{1,n}' \subseteq \mathfrak{A}_{2,n}$, where $\mathfrak{A}'$ is a commutant of a von Neumann algebra $\mathfrak{A}$. We study what happens as $n \to \infty$. In the limit, we obtain some unitary representation, $T^{R,\mu,m}$ (see (3.6)), of the group $G := \mathrm{GL}_0(2\infty, \mathbb{R}) = \varinjlim_{n,i^s} \mathrm{GL}(2n + 1, \mathbb{R})$ acting from the right on X_m. In the generic case, the representation $T^{R,\mu,m}$ is reducible. Indeed, if there exists a non-trivial element

$s \in \mathrm{GL}(m, \mathbb{R})$ such that the left action is *admissible* for the measure $\mu_{(b,a)}^{m}$, i.e., $(\mu_{(b,a)}^{m})^{Ls} \sim \mu_{(b,a)}^{m}$, the operator $T_s^{L,\mu,m}$ naturally associated with the left action, is well defined and $[T_t^{R,\mu,m}, T_s^{L,\mu,m}] = 0$ for all $t \in G$, $s \in \mathrm{GL}(m, \mathbb{R})$.

Here, as in the case of the regular (Kosyak, 1990, 1992) and quasi-regular (Kosyak, 2003) representations of the group $B_0^{\mathbb{N}}$, which is an inductive limit of upper-triangular real matrices, we obtain the remarkable result that the *irreducible representations* can be obtained as the *inductive limit of reducible representations.*

The action of $\mathrm{GL}(2n + 1, \mathbb{R})$ on the space $X_{m,n}$ can be seen as a product of the natural action on $\mathbb{R}^{2n+1}$. To see this, set

$$X_{m,n}^{(k)} = \left\{ x = \sum_{-n \leq r \leq n} x_{kr} E_{kr}, \ x_{kr} \in \mathbb{R} \right\},$$

$$\mu_{(b,a)}^{m,n,k}(x) := \bigotimes_{r=-n}^{n} \mu_{(b_{kr}, a_{kr})}(x_{kr}).$$

Then,

$$X_{m,n} = \bigotimes_{k=1}^{m} \mathbb{R}^{2n+1}, \quad L^2(X_{m,n}, \mu_{(b,a)}^{m,n}) = \bigotimes_{k=1}^{m} L^2(\mathbb{R}^{2n+1}, \mu_{(b,a)}^{m,n,k}).$$

3.2   The Main Result

Let us denote by $\mathrm{Mat}(2\infty, \mathbb{R})$ the space of all real matrices that are infinite in both directions:

$$\mathrm{Mat}(2\infty, \mathbb{R}) = \left\{ x = \sum_{k,n \in \mathbb{Z}} x_{kn} E_{kn}, \ x_{kn} \in \mathbb{R} \right\}. \tag{3.1}$$

The group $\mathrm{GL}_0(2\infty, \mathbb{R}) = \varinjlim_{n, i^s} \mathrm{GL}(2n + 1, \mathbb{R})$ is defined as the inductive limit of the general linear groups $G_n = \mathrm{GL}(2n + 1, \mathbb{R})$ with respect to the *symmetric embedding* i^s:

$$G_n \ni x \mapsto i^s_{n+1}(x) = x + E_{-(n+1),-(n+1)} + E_{n+1,n+1} \in G_{n+1}. \qquad (3.2)$$

For a fixed natural number m, consider a G-space X_m as the following subspace of the space $\mathrm{Mat}(2\infty, \mathbb{R})$:

$$X_m = \left\{ x \in \mathrm{Mat}(2\infty, \mathbb{R}) \mid x = \sum_{k=1}^{m} \sum_{n \in \mathbb{Z}} x_{kn} E_{kn} \right\}. \qquad (3.3)$$

The right action of the group $\mathrm{GL}_0(2\infty, \mathbb{R})$ is correctly defined on the space X_m by the formula $R_t(x) = xt^{-1}$, $t \in G$, $x \in X_m$. We define a Gaussian non-centred product measure, $\mu := \mu^m := \mu^m_{(b,a)}$, on the space X_m:

$$\mu^m_{(b,a)}(x) = \bigotimes_{k=1}^{m} \bigotimes_{n \in \mathbb{Z}} \mu_{(b_{kn}, a_{kn})}(x_{kn}), \qquad (3.4)$$

$$\text{where} \quad d\mu_{(b_{kn}, a_{kn})}(x_{kn}) = \sqrt{\frac{b_{kn}}{\pi}} e^{-b_{kn}(x_{kn} - a_{kn})^2} dx_{kn} \qquad (3.5)$$

and $b = (b_{kn})_{k,n}$, $b_{kn} > 0$, $a = (a_{kn})_{k,n}$, $a_{kn} \in \mathbb{R}$, $1 \leq k \leq m$, $n \in \mathbb{Z}$. Next, we define the unitary representation $T^{R,\mu,m}$ of the group $\mathrm{GL}_0(2\infty, \mathbb{R})$ on the space $L^2(X_m, \mu^m_{(b,a)})$ by the formula

$$(T_t^{R,\mu,m} f)(x) = \left(d\mu^m_{(b,a)}(xt)/d\mu^m_{(b,a)}(x) \right)^{1/2} f(xt), \ f \in L^2(X_m, \mu^m_{(b,a)}). \qquad (3.6)$$

Obviously, the *centraliser* $Z_{\mathrm{Aut}(X_m)}(R(G)) \subset \mathrm{Aut}(X_m)$ contains the group $L(\mathrm{GL}(m, \mathbb{R}))$, i.e., the image of the group $\mathrm{GL}(m, \mathbb{R})$ with respect to the left action $L : \mathrm{GL}(m, \mathbb{R}) \to \mathrm{Aut}(X_m)$, $L_s(x) = sx$, $s \in \mathrm{GL}(m, \mathbb{R})$, $x \in X_m$.

Theorem 3.1. *The representation*

$$T^{R,\mu,m} : \mathrm{GL}_0(2\infty, \mathbb{R}) \to U\left(L^2(X_m, \mu^m_{(b,a)}) \right)$$

is irreducible, for $m = 3$, if and only if

(i) $(\mu^m_{(b,a)})^{L_s} \perp \mu^m_{(b,a)}$ *for all* $s \in \mathrm{GL}(m, \mathbb{R}) \backslash \{e\}$;
(ii) *the measure* $\mu^m_{(b,a)}$ *is G-ergodic.*

In Kosyak (2018, 2019), this result was proved for $m \leq 2$. Note that Theorem 3.1 is a particular case of a generalisation of the Ismagilov conjecture (see Conjecture 7.7 in Kosyak, 2023) for the group G acting on some space X.

Remark 3.1. Any Gaussian product measure $\mu_{(b,a)}^m$ on X_m is $\mathrm{GL}_0(2\infty, \mathbb{R})$-right-ergodic (Shilov and Fan Dik Tun', 1967, Section 3, Corollary 1). For non-product measures, this is not true in general.

In order to study the condition $(\mu_{(b,a)}^m)^{L_t} \perp \mu_{(b,a)}^m$ for $t \in \mathrm{GL}(m, \mathbb{R}) \setminus \{e\}$, set

$$t = (t_{rs})_{r,s=1}^m \in \mathrm{GL}(m, \mathbb{R}), \quad B_n = \mathrm{diag}(b_{1n}, b_{2n}, \ldots, b_{mn}),$$

$$X_n(t) = B_n^{1/2} t B_n^{-1/2}. \tag{3.7}$$

Let $M_{j_1 j_2 \cdots j_r}^{i_1 i_2 \cdots i_r}(t)$ be the *minors* of the matrix t with $i_1, i_2, \ldots, i_r$ rows and $j_1, j_2, \ldots, j_r$ columns, $1 \leq r \leq m$. Let δ_{rs} be the Kronecker symbols.

Lemma 3.1 (Kosyak, 2018, Lemma 10.2.3; Kosyak, 2019, Lemma 2.2). *For the measures* $\mu_{(b,a)}^m$, *with m a natural number, the following relation holds true:*

$$(\mu_{(b,a)}^m)^{L_t} \perp \mu_{(b,a)}^m \text{ for all } t \in \mathrm{GL}(m, \mathbb{R}) \setminus \{e\} \quad \text{if and only if}$$

$$\prod_{n \in \mathbb{Z}} \frac{1}{2^m |\det t|} \det\left(I + X_n^*(t) X_n(t)\right)$$

$$+ \sum_{n \in \mathbb{Z}} \sum_{r=1}^m b_{rn} \left(\sum_{s=1}^m (t_{rs} - \delta_{rs}) a_{sn}\right)^2 = \infty,$$

$$\det(I + X_n^*(t) X_n(t)) = 1 + \sum_{r=1}^m \sum_{\substack{1 \leq i_1 < i_2 < \cdots < i_r \leq m \\ 1 \leq j_1 < j_2 < \cdots < j_r \leq m}} \left(M_{j_1 j_2 \cdots j_r}^{i_1 i_2 \cdots i_r}(X_n(t))\right)^2.$$

$$\tag{3.8}$$

Let us define the following measures on the spaces $\mathbb{R}^m$ and X_m:

$$\mu_m^{(B_n, 0)} = \bigotimes_{k=1}^m \mu_{(b_{kn}, 0)}, \quad \mu_m^{(B_n, a_n)} = \bigotimes_{k=1}^m \mu_{(b_{kn}, a_{kn})},$$

where $a_n = (a_{1n}, \ldots, a_{mn}) \in \mathbb{R}^m$ and $B_n = \mathrm{diag}(b_{1n}, \ldots, b_{mn}) \in \mathrm{Mat}(m, \mathbb{R})$. Set

$$H_{m,n}(t) = H\left(\left(\mu_m^{(B_n,0)}\right)^{L_t}, \mu_m^{(B_n,0)}\right)$$

$$= \left(\frac{1}{2^m |\det t|} \det\left(I + X_n^*(t) X_n(t)\right)\right)^{-1/2}. \tag{3.9}$$

Remark 3.2 (The idea of the proof of irreducibility). Let us denote by $\mathfrak{A}^m$ the *von Neumann algebra* generated by the representation $T^{R,\mu,m}$, i.e., $\mathfrak{A}^m = (T_t^{R,\mu,m} \mid t \in G)''$, where M' is the *commutant* of the von Neumann algebra $M \subset B(H)$. For $\alpha = (\alpha_k) \in \{0,1\}^m$, define the von Neumann algebra $L_\alpha^\infty(X_m, \mu^m)$ as follows:

$$L_\alpha^\infty(X_m, \mu^m) = \left(\exp(it B_{kn}^\alpha) \mid 1 \leq k \leq m, \ t \in \mathbb{R}, \ n \in \mathbb{Z}\right)'',$$

where $B_{kn}^\alpha = \begin{cases} x_{kn}, & \text{if } \alpha_k = 0, \\ i^{-1} D_{kn}, & \text{if } \alpha_k = 1, \end{cases}$ and $D_{kn} = \partial/\partial x_{kn} - b_{kn}(x_{kn} - a_{kn})$.

The proof of the irreducibility is based on four facts:

(1) We can approximate by the generators $A_{kn} = A_{kn}^{R,m} = \frac{d}{dt} T_{I+tE_{kn}}^{R,\mu,m}|_{t=0}$ the set of operators $(B_{kn}^\alpha)_{k=1}^m$, $n \in \mathbb{Z}$, *for some* $\alpha \in \{0,1\}^m$ depending on the measure μ^m using the orthogonality condition $(\mu^m)^{L_s} \perp \mu^m$ for all $s \in \mathrm{GL}(m, \mathbb{R})\backslash\{e\}$.

(2) It is sufficient to verify the approximation only for the *cyclic vector* $\mathbf{1}(x) \equiv 1$ since the representation $T^{R,\mu,m}$ is *cyclic*.

(3) The subalgebra $L_\alpha^\infty(X_m, \mu^m)$ is a *maximal abelian subalgebra* in $\mathfrak{A}^m$.

(4) The measure μ^m is G-ergodic.

Here, the *generators* A_{kn} are given by the formulas

$$A_{kn} = \sum_{r=1}^m x_{rk} D_{rn}, \quad k, n \in \mathbb{Z}, \quad \text{where}$$

$$D_{kn} = \partial/\partial x_{kn} - b_{kn}(x_{kn} - a_{kn}). \tag{3.10}$$

Remark 3.3 (Scheme of the proof). We prove the irreducibility as follows:

$$\left(\mu^{L_s} \perp \mu \ \text{ for all } \ s \in \mathrm{GL}(3,\mathbb{R}) \setminus \{e\}\right) \Leftrightarrow \left(\begin{array}{c} \text{criteria} \\ \text{of} \\ \text{orthogonality} \end{array}\right) \ \& \quad (3.11)$$

$$\left(\begin{array}{c} \text{Lemma 6.4} \\ \text{about} \\ \text{three vectors } f,g,h \notin l_2 \end{array}\right) \Rightarrow \left(\begin{array}{c} \text{some of} \quad \Delta^{(1)}, \Delta_1 \\ \text{the expressions } \Delta^{(2)}, \Delta_2 \\ \text{are divergent:} \ \ \Delta^{(3)}, \Delta_3 \end{array}\right) \Rightarrow \text{irreducibility},$$

$$\text{where} \quad \Delta^{(i)} := \Delta(Y_i^{(i)}, Y_j^{(i)}, Y_k^{(i)}), \quad \Delta_i := \Delta(Y_i, Y_j, Y_k), \quad (3.12)$$

$\Delta(f,g,h)$ is defined by (3.15), and $\{i,j,k\}$ is a cyclic permutation of $\{1,2,3\}$; see Lemmas 5.1, 5.2 and 5.4 for details.

We use the following notation: for k vectors $f_1, f_2, \ldots, f_k \in \mathbb{R}^n$, with $k \leq n$, set

$$\Delta(f_1, f_2, \ldots, f_k) = \frac{\det(I + \gamma(f_1, f_2, \ldots, f_k))}{\det(I + \gamma(f_2, \ldots, f_k))} - 1 \quad \text{for } k = 2,3, \quad (3.13)$$

$$\Delta(f_1, f_2) = \frac{\det(I + \gamma(f_1, f_2))}{\det(I + \gamma(f_2))} - 1$$
$$= \frac{I + \Gamma(f_1) + \Gamma(f_2) + \Gamma(f_1, f_2)}{I + \Gamma(f_2)}, \quad (3.14)$$

$$\Delta(f_1, f_2, f_3) = \frac{\Gamma(f_1) + \Gamma(f_1, f_2) + \Gamma(f_1, f_3) + \Gamma(f_1, f_2, f_3)}{1 + \Gamma(f_2) + \Gamma(f_3) + \Gamma(f_2, f_3)}.$$
$$(3.15)$$

Remark 3.4. The fact that the conditions $(\mu^3_{(b,a)})^{L_t} \perp \mu^3_{(b,a)}$ for all $t \in \mathrm{GL}(3,\mathbb{R}) \setminus \{e\}$ imply the possibility of the approximation of x_{kn} or D_{kn} by combinations of generators is based on Lemmas 6.1 and 6.5 about the explicit expression for $(C^{-1}(\lambda)a, a)$ (see Kosyak, 2023), where $C(\lambda)$ is defined by (7.6). Finally, the last lemma is based on some completely *independent statement* about three infinite vectors, $f_1, f_2, f_3 \notin l_2(\mathbb{N})$, such that $\sum_{k=1}^{3} C_k f_k \notin l_2(\mathbb{N})$; see Lemma 6.4 for general m in Kosyak (2025a). These lemmas are the *key ingredients* for the proof of the irreducibility of the representation.

Remark 3.5. Note that in the case of the "nilpotent group" $B_0^{\mathbb{N}}$ and the infinite product of *arbitrary* Gaussian measures on $\mathbb{R}^m$ (see Albeverio and Kosyak, 2006), the proof of the *irreducibility* is also based on another completely *independent statement*, namely, *the Hadamard–Fischer inequality*; see Lemma 3.2.

Lemma 3.2 (Hadamard–Fischer inequality, Horn and Johnson, 1989, 1991). *For any positive definite matrix* $C \in \mathrm{Mat}(m, \mathbb{R})$, $m \in \mathbb{N}$, *and any two subsets* α *and* β *with* $\emptyset \subseteq \alpha$, $\beta \subseteq \{1, \dots, m\}$, *the following inequality holds:*

$$\begin{vmatrix} M(\alpha) & M(\alpha \bigcap \beta) \\ M(\alpha \bigcup \beta) & M(\beta) \end{vmatrix} = \begin{vmatrix} A(\hat{\alpha}) & A(\hat{\alpha} \bigcup \hat{\beta}) \\ A(\hat{\alpha} \bigcap \hat{\beta}) & A(\hat{\beta}) \end{vmatrix} \geq 0, \qquad (3.16)$$

where $M(\alpha) = M_\alpha^\alpha(C)$, $A(\alpha) = A_\alpha^\alpha(C)$ *are factors and cofactors of* C *and* $\hat{\alpha} = \{1, \dots, m\} \setminus \alpha$.

3.2.1 *Equivalent series and equivalent sequences*

Definition 3.1. We say that two series $\sum_{n \in \mathbb{N}} a_n$ and $\sum_{n \in \mathbb{N}} b_n$ with positive a_n, b_n are *equivalent* if they are divergent or convergent simultaneously. We denote it as $\sum_{n \in \mathbb{N}} a_n \sim \sum_{n \in \mathbb{N}} b_n$. We say that two sequences $(a_n)_{n \in \mathbb{N}}$ and $(b_n)_{n \in \mathbb{N}}$ are *equivalent* if for some $C_1, C_2 > 0$, we have $C_1 b_n \leq a_n \leq C_2 b_n$ for all $n \in \mathbb{N}$. We use the same notation: $a_n \sim b_n$.

Lemma 3.3. *Let* $1 + c_n > 0$ *for all* $n \in \mathbb{Z}$. *Then, the following two series are equivalent:*

$$\Sigma_1 := \sum_{n \in \mathbb{Z}} \frac{c_n^2}{1 + c_n}, \quad \Sigma_2 := \sum_{n \in \mathbb{Z}} c_n^2. \qquad (3.17)$$

Proof. Fix some $\varepsilon \in (0, 1)$ and a large N. We have three cases:

(a) $1 + c_n \in (\varepsilon, N)$;
(b) for the infinite subset $\mathbb{Z}_1$, we have $\lim_{n \in \mathbb{Z}_1} c_n = \infty$;
(c) for the infinite subset $\mathbb{Z}_1$, we have $\lim_{n \in \mathbb{Z}_1} (1 + c_n) = 0$.

When Case (a) is not applicable, there is an infinite subset $\mathbb{Z}_1 \subset \mathbb{Z}$ such that (b) or (c) holds. In Case (a), we have

$$\frac{1}{N} \sum_{n \in \mathbb{Z}} c_n^2 < \sum_{n \in \mathbb{Z}} \frac{c_n^2}{1 + c_n} < \frac{1}{\varepsilon} \sum_{n \in \mathbb{Z}} c_n^2. \tag{3.18}$$

In Cases (b) and (c), both series are divergent. $\qquad\qquad\square$

We will make systematic use of the following statement.

Remark 3.6 (Kosyak, 2018). Let $a_n, b_n > 0$ for all $n \in \mathbb{N}$. The following two series are equivalent:

$$\sum_{n \in \mathbb{N}} \frac{a_n}{a_n + b_n} \sim \sum_{n \in \mathbb{N}} \frac{a_n}{b_n}. \tag{3.19}$$

3.3 Some Orthogonality Problems in Measure Theory

3.3.1 *General setting*

Our aim now is to find the minimal generating set of conditions for the orthogonality $(\mu_{(b,a)}^m)^{L_t} \perp \mu_{(b,a)}^m$ for all $t \in \mathrm{GL}(m, \mathbb{R}) \setminus \{e\}$. To be more precise, consider the following more general situation. Let $\alpha : G \to \mathrm{Aut}(X)$ be a *measurable action* of a group G on a measurable space (X, μ) with the following property: $\mu^{\alpha_t} \perp \mu$ for all $t \in G \setminus \{e\}$. Define a *generating subset* $G^\perp(\mu)$ in the group G as follows:

$$\text{if}\quad \mu^{\alpha_t} \perp \mu \text{ for all } t \in G^\perp(\mu), \quad \text{then}\quad \mu^{\alpha_t} \perp \mu \text{ for all } t \in G \setminus \{e\}.$$
$$\tag{3.20}$$

Problem 3.1. Find a minimal generating subset $G_0^\perp(\mu)$ satisfying (3.20).

3.3.2 *Orthogonality criteria $\mu^{L_t} \perp \mu$ for $t \in \mathrm{GL}(2, \mathbb{R}) \setminus \{e\}$*

Remark 3.7. By Lemma 4.1 proved in Kosyak (2019) or Lemma 10.4.1 in Kosyak (2018) for $m = 2$, we conclude that the minimal generating set $G_0^{\perp}(\mu) = \mathrm{GL}(2, \mathbb{R})_0^{\perp}(\mu)$ (see Problem 3.1) is reduced to the following subgroups, families and elements:

$$\exp(tE_{12}) = I + tE_{12} = \begin{pmatrix} 1 & t \\ 0 & 1 \end{pmatrix}, \quad \exp(tE_{21}) = I + tE_{21} = \begin{pmatrix} 1 & 0 \\ t & 1 \end{pmatrix},$$

$$(3.21)$$

$$\exp(tE_{12})P_1 = \begin{pmatrix} -1 & t \\ 0 & 1 \end{pmatrix}, \quad \exp(tE_{21})P_2 = \begin{pmatrix} 1 & 0 \\ t & -1 \end{pmatrix}, \qquad (3.22)$$

$$\tau_-(\phi, s) = \begin{pmatrix} \cos\phi & s^2 \sin\phi \\ s^{-2} \sin\phi & -\cos\phi \end{pmatrix} = D_2(s) \begin{pmatrix} \cos\phi & -\sin\phi \\ \sin\phi & \cos\phi \end{pmatrix} D_2^{-1}(s)P_2.$$

$$(3.23)$$

The families (3.21) are one-parameter subgroups, the family (3.22) are simply reflections of (3.21) and the family (3.23) depends on two parameters. All elements are of order 2 except the elements in subgroups given in (3.21). It suffices to verify the conditions (3.21) only for some $t \in \mathbb{R} \setminus \{0\}$. The family $\tau_-(\phi, s)$, in fact, coincides with $D_2(s)O(2)D_2^{-1}(s)P_2$, where $D_2(s) = \mathrm{diag}(s, s^{-1})$. All points t in (3.22) and all points (ϕ, s) in (3.23) are essential, i.e., we cannot remove any of the points.

The conditions for the orthogonality with respect to elements defined by (3.21)–(3.23) are transformed in the divergence of the following series:

$$S_{12}^L(\mu) = \sum_{n \in \mathbb{Z}} \frac{b_{1n}}{2} \left(\frac{1}{2b_{2n}} + a_{2n}^2 \right), \quad S_{21}^L(\mu) = \sum_{n \in \mathbb{Z}} \frac{b_{2n}}{2} \left(\frac{1}{2b_{1n}} + a_{1n}^2 \right);$$

$$(3.24)$$

$$S_{kn}^{L,-}(\mu, t) = \frac{t^2}{4} \sum_{m \in \mathbb{Z}} \frac{b_{km}}{b_{nm}} + \sum_{m \in \mathbb{Z}} \frac{b_{km}}{2}(-2a_{km} + ta_{nm})^2, \quad t \in \mathbb{R};$$

$$(3.25)$$

$$\Sigma_{12}^-(\tau_-(\phi, s)) = \sin^2\phi \, \Sigma_{12}(s) + \Sigma_{12}^-(\tau_-(\phi, s)), \quad \phi \in [0, 2\pi), \quad s > 0,$$

$$(3.26)$$

$$\text{where} \quad \Sigma_{12}(s) := \sum_{n \in \mathbb{Z}} \left(s^2 \sqrt{\frac{b_{1n}}{b_{2n}}} - s^{-2} \sqrt{\frac{b_{2n}}{b_{1n}}} \right)^2 ; \tag{3.27}$$

$$\Sigma_{12}^{-}\big(\tau_{-}(\phi, s)\big) = \sum_{n \in \mathbb{Z}} \left(4b_{1n} \sin^2 \frac{\phi}{2} + 4s^{-4} b_{2n} \cos^2 \frac{\phi}{2} \right)$$

$$\times \left(a_{1n} \sin \frac{\phi}{2} - s^2 a_{2n} \cos \frac{\phi}{2} \right)^2 . \tag{3.28}$$

Remark 3.8 (Kosyak, 2019). The following three conditions are equivalent:

(i) $\mu^{L_{\tau_{-}(\phi,s)}} \perp \mu, \quad \phi \in [0, 2\pi), \ s > 0;$

(ii) $\Sigma_{12}\big(\tau_{-}(\phi,s)\big) = \sin^2 \phi \, \Sigma_{12}(s) + \Sigma_{12}^{-}\big(\tau_{-}(\phi, s)\big) = \infty,$
      $\phi \in [0, 2\pi), \ s > 0;$

(iii) $\Sigma_{12}(s) + \Sigma_{12}(C_1, C_2) = \infty, \quad s > 0, \ (C_1, C_2) \in \mathbb{R}^2 \setminus \{0\},$

where $\Sigma_{12}(s)$ is defined by (3.27) and

$$\Sigma_{12}(C_1, C_2) := \sum_{n \in \mathbb{Z}} (C_1^2 b_{1n} + C_2^2 b_{2n})(C_1 a_{1n} + C_2 a_{2n})^2. \tag{3.29}$$

3.3.3   *Orthogonality criteria $\mu^{L_t} \perp \mu$ for $t \in \mathbf{GL}(3, \mathbb{R}) \setminus \{e\}$*

Recall (Kosyak, 2019) that for $m = 2$ and $\det t > 0$, we have

$$2^2 |\det t| \big(H_{2,n}^{-2}(t) - 1 \big)$$

$$= \left[(1 - |\det t|)^2 + (t_{11} - t_{22})^2 + \left(t_{12} \sqrt{\frac{b_{1n}}{b_{2n}}} + t_{21} \sqrt{\frac{b_{2n}}{b_{1n}}} \right)^2 \right]$$

$$= \Big[\big(M_{\emptyset}^{\emptyset}(X(t)) - A_{\emptyset}^{\emptyset}(X(t)) \big)^2 + \big(M_1^1(X(t)) - A_1^1(X(t)) \big)^2$$

$$+ \big(M_2^1(X(t)) - A_2^1(X(t)) \big)^2 \Big],$$

where $H_{m,n}(t)$ is defined by (3.9). For $m = 3$, using (3.7), we have $X(t) = B^{1/2}tB^{-1/2}$, and hence

$$X(t) = \begin{pmatrix} b_{1n} & 0 & 0 \\ 0 & b_{2n} & 0 \\ 0 & 0 & b_{3n} \end{pmatrix}^{1/2} \begin{pmatrix} t_{11} & t_{12} & t_{13} \\ t_{21} & t_{22} & t_{23} \\ t_{31} & t_{32} & t_{33} \end{pmatrix} \begin{pmatrix} b_{1n} & 0 & 0 \\ 0 & b_{2n} & 0 \\ 0 & 0 & b_{3n} \end{pmatrix}^{-1/2}$$

$$= \begin{pmatrix} t_{11} & \sqrt{\frac{b_{1n}}{b_{2n}}}\, t_{12} & \sqrt{\frac{b_{1n}}{b_{3n}}}\, t_{13} \\ \sqrt{\frac{b_{2n}}{b_{1n}}}\, t_{21} & t_{22} & \sqrt{\frac{b_{2n}}{b_{3n}}}\, t_{23} \\ \sqrt{\frac{b_{3n}}{b_{1n}}}\, t_{31} & \sqrt{\frac{b_{3n}}{b_{2n}}}\, t_{32} & t_{33} \end{pmatrix}.$$

Therefore, using (3.8) and the fact that $X = X^*(t)X(t)$, we obtain

$$2^3 |\det t| H_{3,n}^{-2}(t)$$

$$= \left(1 + |\det t|^2 + t_{11}^2 + \frac{b_{1n}}{b_{2n}} t_{12}^2 + \frac{b_{1n}}{b_{3n}} t_{13}^2 \right.$$

$$+ \frac{b_{2n}}{b_{1n}} t_{21}^2 + t_{22}^2 + \frac{b_{2n}}{b_{3n}} t_{23}^2 + \frac{b_{3n}}{b_{1n}} t_{31}^2 + \frac{b_{3n}}{b_{2n}} t_{32}^2 + t_{33}^2 + \left(M_{12}^{12}(t) \right)^2$$

$$+ \frac{b_{2n}}{b_{3n}} \left(M_{13}^{12}(t) \right)^2 + \frac{b_{1n}}{b_{3n}} \left(M_{23}^{12}(t) \right)^2 + \frac{b_{3n}}{b_{2n}}$$

$$\times \left(M_{12}^{13}(t) \right)^2 + \left(M_{13}^{13}(t) \right)^2 + \frac{b_{1n}}{b_{2n}} \left(M_{23}^{13}(t) \right)^2 + \frac{b_{3n}}{b_{1n}} \left(M_{12}^{23}(t) \right)^2 + \frac{b_{2n}}{b_{2n}}$$

$$\left. \times \left(M_{13}^{23}(t) \right)^2 + \left(M_{23}^{23}(t) \right)^2 \right)$$

$$= 1 + |\det t|^2 + \sum_{1 \le i \le j \le 3} \left[\left(t_j^i \sqrt{\frac{b_{in}}{b_{jn}}} \right)^2 + \left(A_j^i \sqrt{\frac{b_{jn}}{b_{in}}} \right)^2 \right]$$

$$= 1 + |\det t|^2 + \sum_{1 \le i \le j \le 3} \left(|M_j^i(X(t))|^2 + |A_j^i(X(t))|^2 \right).$$

Let $A_j^i(t)$ be the *cofactors* of a matrix $t \in \mathrm{GL}(3, \mathbb{R})$. Using the notation $t_j^i = t_{ij}$ and the relation

$$\det t = t_1^k A_1^k(t) + t_2^k A_2^k(t) + t_3^k A_3^k(t), \quad k = 1, 2, 3, \quad \text{we get}$$

$$2^3|\det t|(H_{3,n}^{-2}(t) - 1)$$

$$= (1 - |\det t|)^2 + \sum_{1 \le i,j \le 3} \left(M_j^i(X(t)) - A_j^i(X(t)) \right)^2$$

$$= (1 - |\det t|)^2 + \sum_{1 \le i \le j \le 3} \left(t_j^i \sqrt{\frac{b_{in}}{b_{jn}}} - A_j^i(t) \sqrt{\frac{b_{jn}}{b_{in}}} \right)^2. \tag{3.30}$$

Similar to Kosyak (2019, Lemmas 2.22) in the case of $m = 2$ or Kosyak (2018, Lemma 10.4.30), we get the following lemma for $m = 3$.

Lemma 3.4. *For $t \in \mathrm{GL}(3, \mathbb{R}) \setminus \{e\}$, if $\pm \det t > 0$, we have, respectively,*

$$(\mu_{(b,0)}^3)^{L_t} \perp \mu_{(b,0)}^3$$

$$\Leftrightarrow \sum_{n \in \mathbb{Z}} \left[(1 - |\det t|)^2 + \sum_{1 \le i \le 3} \left(t_i^i \mp A_i^i(t) \right)^2 \right.$$

$$\left. + \sum_{1 \le i < j \le 3} \left(t_j^i \sqrt{\frac{b_{in}}{b_{jn}}} \mp A_j^i(t) \sqrt{\frac{b_{jn}}{b_{in}}} \right)^2 \right] = \infty. \tag{3.31}$$

By Lemma 3.1, the following lemma holds true.

Lemma 3.5. *For $t \in \mathrm{GL}(3, \mathbb{R}) \setminus \{e\}$, we have*

$$(\mu_{(b,a)}^3)^{L_t} \perp \mu_{(b,a)}^3 \quad \text{if} \quad |\det t| \ne 1.$$

If $\det t = \pm 1$, we have, respectively,

$$(\mu_{(b,a)}^3)^{L_t} \perp \mu_{(b,a)}^3 \quad \Leftrightarrow \quad \Sigma^{\pm}(t) := \Sigma_1^{\pm}(t) + \Sigma_2(t) = \infty, \quad \text{where}$$

$$\Sigma_1^+(t) = \sum_{n \in \mathbb{Z}} \left[\sum_{k=1}^3 (t_{kk} - A_k^k(t))^2 + \sum_{1 \le i < j \le 3} \left(t_j^i \sqrt{\frac{b_{in}}{b_{jn}}} - A_j^i(t) \sqrt{\frac{b_{jn}}{b_{in}}} \right)^2 \right],$$

$$\tag{3.32}$$

$$\Sigma_1^-(t) = \sum_{n \in \mathbb{Z}} \left[\sum_{k=1}^3 (t_{kk} + A_k^k(t))^2 + \sum_{1 \le i < j \le 3} \left(t_j^i \sqrt{\frac{b_{in}}{b_{jn}}} + A_j^i(t) \sqrt{\frac{b_{jn}}{b_{in}}} \right)^2 \right],$$

$$\tag{3.33}$$

$$\Sigma_2(t^{-1}) = \sum_{n\in\mathbb{Z}} \Big[b_{1n}\big((t_{11}-1)a_{1n} + t_{12}a_{2n} + t_{13}a_{3n}\big)^2$$

$$+ \, b_{2n}\big(t_{21}a_{1n} + (t_{22}-1)a_{2n} + t_{23}a_{3n}\big)^2$$

$$+ \, b_{3n}\big(t_{31}a_{1n} + t_{32}a_{2n} + (t_{33}-1)a_{3n}\big)^2\Big]. \qquad (3.34)$$

Remark 3.9. By Lemma 3.5, it suffices to verify the condition of orthogonality

$$(\mu^3_{(b,a)})^{L_t} \perp \mu^3_{(b,a)} \quad \text{for} \quad t \in \mathrm{GL}(3,\mathbb{R}) \setminus \{e\},$$

for the following two subsets of the group $\pm\mathrm{SL}(3,\mathbb{R})$:

$$G_3^{\pm} := \{ t \in \pm\mathrm{SL}(3,\mathbb{R}) \mid t_{kk} = \pm A_k^k(t),\ 1 \le k \le 3 \}. \qquad (3.35)$$

Lemma 3.6. *If $t \in G_3^{\pm}$, we have, respectively,*

$$(\mu^3_{(b,a)})^{L_t} \perp \mu^3_{(b,a)} \quad \Leftrightarrow \quad \Sigma^{\pm}(t) = \Sigma_1^{\pm}(t) + \Sigma_2(t) = \infty,$$

$$\Sigma_1^{\pm}(t) = \sum_{1\le i<j\le 3}\sum_{n\in\mathbb{Z}} \left(t_j^i \sqrt{\frac{b_{in}}{b_{jn}}} \mp A_j^i(t)\sqrt{\frac{b_{jn}}{b_{in}}} \right)^2 = \sum_{1\le i<j\le 3} \Sigma_{ij}^{\pm}(t),$$

$$(3.36)$$

$$\Sigma_{ij}^{\pm}(t) = \sum_{n\in\mathbb{Z}} \left(t_j^i \sqrt{\frac{b_{in}}{b_{jn}}} \mp A_j^i(t)\sqrt{\frac{b_{jn}}{b_{in}}} \right)^2, \qquad (3.37)$$

where $\Sigma_2(t)$ is defined by (3.34).

Next, we show that the set G_3^{+} can be reduced to the six families of one-parameter subgroups $\exp(tE_{kr})$, $1 \le k \ne r \le 3$ (see (3.39)) or to three families of two-parameter subgroups (see (3.40)). The set G_3^{-} can be reduced to the three two-parameter families (3.41), which are reflections of (3.40) by P_r. The remaining part is reduced to the sets

$D_3(s)O(3)D_3^{-1}(s)P_r$ or the five-parameter family of elements $\tau_r(t,s) = D_3(s)tD_3^{-1}(s)P_r$ (see (3.45)).

Lemma 3.7. *In the case of* $m = 3$, *the* minimal generating set $\mathrm{GL}(3,\mathbb{R})_0^\perp(\mu)$ *is defined as follows (compare with Remark 3.7):*

$$\mathrm{GL}(3,\mathbb{R})_0^\perp(\mu) = \{e_r(t,s), e_r(t,s)P_r, \mid 1 \le r \le 3, \ (t,s) \in \mathbb{R}^2\}$$

$$\cup \{O_r^A(3), \ 1 \le r \le 3\}, \tag{3.38}$$

where $\quad e_{kn}(t) := \exp(tE_{kn}) = I + tE_{kn}, \ 1 \le k \ne n \le 3, \ t \in \mathbb{R},$

$$\tag{3.39}$$

$$e_1(t,s) = \begin{pmatrix} 1 & t & s \\ 0 & 1 & 0 \\ 0 & 0 & 1 \end{pmatrix}, \quad e_2(t,s) = \begin{pmatrix} 1 & 0 & 0 \\ t & 1 & s \\ 0 & 0 & 1 \end{pmatrix}, \quad e_3(t,s) = \begin{pmatrix} 1 & 0 & 0 \\ 0 & 1 & 0 \\ t & s & 1 \end{pmatrix}, \tag{3.40}$$

$$e_1(t,s)P_1 = \begin{pmatrix} -1 & t & s \\ 0 & 1 & 0 \\ 0 & 0 & 1 \end{pmatrix}, \quad e_2(t,s)P_2 = \begin{pmatrix} 1 & 0 & 0 \\ t & -1 & s \\ 0 & 0 & 1 \end{pmatrix},$$

$$e_3(t,s)P_3 = \begin{pmatrix} 1 & 0 & 0 \\ 0 & 1 & 0 \\ t & s & -1 \end{pmatrix}, \tag{3.41}$$

$$P_1 = \begin{pmatrix} -1 & 0 & 0 \\ 0 & 1 & 0 \\ 0 & 0 & 1 \end{pmatrix}, \quad P_2 = \begin{pmatrix} 1 & 0 & 0 \\ 0 & -1 & 0 \\ 0 & 0 & 1 \end{pmatrix}, \quad P_3 = \begin{pmatrix} 1 & 0 & 0 \\ 0 & 1 & 0 \\ 0 & 0 & -1 \end{pmatrix}, \tag{3.42}$$

$$O^A(3) := \{D_3(s)O(3)D_3^{-1}(s) \mid D_3(s) \in A\}, \tag{3.43}$$

$$O_r^A(3) := \{D_3(s)O(3)D_3^{-1}(s)P_r \mid D_3(s) \in A\}, \ 1 \le r \le 3, \tag{3.44}$$

$$\tau_r(t,s) = D_3(s)tD_3^{-1}(s)P_r, \ t \in O(3),$$

$$A = \{D_3(s) = \mathrm{diag}(s_1, s_2, s_3)\}. \tag{3.45}$$

The families (3.39) give us, respectively, the divergence of the following series:

$$S_{kr}^L(\mu) = \sum_{n \in \mathbb{Z}} \frac{b_{kn}}{2}\left(\frac{1}{2b_{rn}} + a_{rn}^2\right), \quad 1 \le k, r \le 3, \ k \ne r. \tag{3.46}$$

The families (3.40) give us the divergence of the following series:

$$S_{1,23}^L(\mu, t, s) = \sum_{n \in \mathbb{Z}}\left[\frac{t^2}{4}\frac{b_{1n}}{b_{2n}} + \frac{s^2}{4}\frac{b_{1n}}{b_{3n}} + \frac{b_{1n}}{2}(-2a_{1n} + ta_{2n} + sa_{3n})^2\right],$$

$$\tag{3.47}$$

$$S_{2,31}^{L}(\mu, t, s) = \sum_{n \in \mathbb{Z}} \left[\frac{t^2}{4} \frac{b_{2n}}{b_{1n}} + \frac{s^2}{4} \frac{b_{2n}}{b_{3n}} + \frac{b_{2n}}{2} \left(t a_{1n} - 2 a_{2n} + s a_{3n} \right)^2 \right],$$

$$(3.48)$$

$$S_{3,12}^{L}(\mu, t, s) = \sum_{n \in \mathbb{Z}} \left[\frac{t^2}{4} \frac{b_{3n}}{b_{1n}} + \frac{s^2}{4} \frac{b_{3n}}{b_{2n}} + \frac{b_{3n}}{2} \left(t a_{1n} + s a_{2n} - 2 a_{3n} \right)^2 \right],$$

$$(3.49)$$

$$\text{in particular,} \quad S_{rr}^{L}(\mu) := S_{r,st}^{L}(\mu, 0, 0) = \sum_{n \in \mathbb{Z}} \frac{b_{rn}}{2} a_{rn}^2, \qquad (3.50)$$

where (r, s, t) is a cyclic permutation of $(1, 2, 3)$. The families (3.45) give us the condition (3.53); see Lemma 3.8 as follows.

Proof. Consider the subset $\mathrm{GL}(3, \mathbb{R})_0^{\perp}(\mu)$ of $\mathrm{GL}(3, \mathbb{R})$, described by (3.38). The fact that this set is *minimal generating* will follow from Lemma 4.1, or, more precisely, from the following implications:

$$\left(\mu^{L_t} \perp \mu \text{ for all } t \in \mathrm{GL}(3, \mathbb{R})_0^{\perp}(\mu) \right) \Rightarrow \left(\text{irreducibility} \right)$$

$$\Rightarrow \left(\mu^{L_t} \perp \mu \text{ for all } t \in \mathrm{GL}(3, \mathbb{R}) \setminus \{e\} \right). \qquad (3.51)$$

The first implication follows from Lemma 4.1, and the second follows from irreducibility. Indeed, suppose that $\mathrm{GL}(3, \mathbb{R})_0^{\perp}(\mu)$ is not a minimal generating set. Then, we can find an $s \in \mathrm{GL}(3, \mathbb{R}) \setminus \{e\}$ such that

$$\left(\mu_{(b,a)}^3 \right)^{L_s} \sim \mu_{(b,a)}^3.$$

Hence, the non-trivial operator $T_s^{L,\mu,3}$ can be defined by

$$(T_s^{L,\mu,3} f)(x) = \left(d\mu_{(b,a)}^3 (s^{-1} x) / d\mu_{(b,a)}^3 (x) \right)^{1/2} f(s^{-1} x), \ f \in L^2(X_3, \mu_{(b,a)}^3).$$

$$(3.52)$$

This operator commutes with the representations $T^{R,\mu,3}$:

$$[T_t^{R,\mu,3}, T_s^{L,\mu,3}] = 0 \quad \text{for all} \quad t \in G,$$

contradicting irreducibility. The relations (3.46)–(3.49) follow from (3.32)–(3.34). The relation (3.48), for example, follows from (3.33) and (3.34). The relation (3.53) is obtained from (3.32) for $\tau_r(t, s)$, $t \in O(3)$, $s \in \left(\mathbb{R}^* \right)^3$, defined by (3.45). $\qquad \square$

Lemma 3.8. *Set* $\tau(s,t) := D_3(s)tD_3^{-1}(s)$ *and* $\tau_r(s,t) := \tau(s,t)P_r$ *for* $t \in \pm O(3)$, $D_3(s) = \mathrm{diag}(s_1, s_2, s_3)$, $s = (s_1, s_2, s_3) \in (\mathbb{R}^*)^3$ *and* $1 \leq r \leq 3$. *Then,*

$$\left(\mu_{(b,a)}^3\right)^{L_{\tau_r(s,t)}} \perp \mu_{(b,a)}^3 \Leftrightarrow \Sigma_1^{\pm}\big(\tau_r(s,t)\big) + \Sigma_2\big(\tau_r(s,t)\big) = \infty, \qquad (3.53)$$

where $\Sigma_1^{\pm}(t)$ *are defined by* (3.36) *and* $\Sigma_2(t)$ *is defined by* (3.34). *In particular, if we denote* $s_{ij} = s_i s_j^{-1}$, *we get*

$$\Sigma_1^{+}\big(\tau(t,s)\big) = \Sigma_1^{+}(t,s)$$
$$= t_{12}^2 \Sigma_{12}(s_{12}^{1/2}) + t_{13}^2 \Sigma_{13}(s_{13}^{1/2}) + t_{23}^2 \Sigma_{23}(s_{23}^{1/2}). \qquad (3.54)$$

Proof.   For $T := \tau(s,t)$ and $T(3) := \tau_3(s,t)$, we have, respectively,

$$T = D_3(s)tD_3^{-1}(s) = \begin{pmatrix} t_{11} & \frac{s_1}{s_2}t_{12} & \frac{s_1}{s_3}t_{13} \\ \frac{s_2}{s_1}t_{21} & t_{22} & \frac{s_2}{s_3}t_{23} \\ \frac{s_3}{s_1}t_{31} & \frac{s_3}{s_2}t_{32} & t_{33} \end{pmatrix}, \qquad (3.55)$$

$$\begin{pmatrix} t_{11} & \frac{s_1}{s_2}t_{12} & -\frac{s_1}{s_3}t_{13} \\ \frac{s_2}{s_1}t_{21} & t_{22} & -\frac{s_2}{s_3}t_{23} \\ \frac{s_3}{s_1}t_{31} & \frac{s_3}{s_2}t_{32} & -t_{33} \end{pmatrix} = D_3(s)tD_3^{-1}(s)P_3 =: T(3). \qquad (3.56)$$

By Lemma 3.9 for $t \in O(3)$, we have $t_{kr} = A_r^k(t)$, $1 \leq k, r \leq 3$. Therefore, for T and $T(3)$, we have for $1 \leq k$ and $r \leq 3$,

$$M_r^k(T) = T_{kr} = \frac{s_k}{s_r}t_{kr}, \quad A_r^k(T) = \frac{s_r}{s_k}A_r^k(t) \overset{(3.59)}{=} \frac{s_r}{s_k}t_{kr}, \qquad (3.57)$$

$$M_r^k(T(3)) = (-1)^{\delta_{3,r}}\frac{s_k}{s_r}t_{kr}, \quad A_r^k(T(3)) = (-1)^{\delta_{3,r}}\frac{s_r}{s_k}A_r^k(t)$$

$$= (-1)^{\delta_{3,r}}\frac{s_r}{s_k}t_{kr}. \qquad (3.58)$$

Finally, we get

$$\Sigma_1^+(T) = \Sigma_1^+\big(\tau(s,t)\big)$$

$$= \sum_{n\in\mathbb{Z}}\left[\sum_{1\le i<j\le 3}\left(M_j^i(T)\sqrt{\frac{b_{in}}{b_{jn}}} - A_j^i(T)\sqrt{\frac{b_{jn}}{b_{in}}}\right)^2\right]$$

$$= \sum_{n\in\mathbb{Z}}\left[t_{12}^2\left(s_{12}\sqrt{\frac{b_{1n}}{b_{2n}}} - s_{12}^{-1}\sqrt{\frac{b_{2n}}{b_{1n}}}\right)^2 + t_{13}^2\left(s_{13}\sqrt{\frac{b_{1n}}{b_{3n}}} - s_{13}^{-1}\sqrt{\frac{b_{3n}}{b_{1n}}}\right)^2\right.$$

$$\left. + t_{23}^2\left(s_{23}\sqrt{\frac{b_{2n}}{b_{3n}}} - s_{23}^{-1}\sqrt{\frac{b_{3n}}{b_{2n}}}\right)^2\right]$$

$$= t_{12}^2\Sigma_{12}(s_{12}^{1/2}) + t_{13}^2\Sigma_{13}(s_{13}^{1/2}) + t_{23}^2\Sigma_{23}(s_{23}^{1/2}).$$

Hence, $\Sigma_1^-(T(3)) = t_{12}^2\Sigma_{12}(s_{12}^{1/2}) + t_{13}^2\Sigma_{13}(s_{13}^{1/2}) + t_{23}^2\Sigma_{23}(s_{23}^{1/2}).$ $\square$

Lemma 3.9. *For an arbitrary orthogonal matrix* $t \in \pm O(3)$, *we have*

$$t_{kn} = \pm A_n^k(t), \ 1 \le k,n \le 3, \quad where \quad t = \begin{pmatrix} t_{11} & t_{12} & t_{13} \\ t_{21} & t_{22} & t_{23} \\ t_{31} & t_{32} & t_{33} \end{pmatrix}. \tag{3.59}$$

Proof. Denote the three rows of the matrix t by, respectively, $t_1, t_2, t_3 \in \mathbb{R}^3$. Since $t \in \pm O(3)$, we get

$$\|t_1\|^2 = \|t_2\|^2 = \|t_3\|^2 = 1 \quad \text{and} \quad t_l \perp t_r, \ l \ne r. \tag{3.60}$$

Moreover, since t_1 is orthogonal to the hyperplane V_{23} generated by the vectors t_2 and t_3, and $t \in \pm O(3)$, we get, respectively, $t_l = \pm[t_r, t_s]$, where $[x, y]$ is the *vector product* or *cross product* of the two vectors $x, y \in \mathbb{R}^3$ and the triple $\{l, r, s\}$ denotes any cyclic permutations of $\{1, 2, 3\}$. For $t \in O(3)$ and $l = 1$, we get

$$t_1 = [t_2, t_3] = \begin{vmatrix} i & j & k \\ t_{21} & t_{22} & t_{23} \\ t_{31} & t_{32} & t_{33} \end{vmatrix} = i\begin{vmatrix} t_{22} & t_{23} \\ t_{32} & t_{33} \end{vmatrix} - j\begin{vmatrix} t_{21} & t_{23} \\ t_{31} & t_{33} \end{vmatrix} + k\begin{vmatrix} t_{21} & t_{22} \\ t_{31} & t_{32} \end{vmatrix}, \tag{3.61}$$

where i, j, k is the standard orthonormal basis in $\mathbb{R}^3$, i.e.,

$$i = (1,0,0), \ \ j = (0,1,0), \ \ k = (0,0,1).$$

Define X *formally* as the matrix

$$X = \begin{pmatrix} i & j & k \\ x_1 & x_2 & x_3 \\ y_1 & y_2 & y_3 \end{pmatrix}, \text{ then } t_1 = (t_{11}, t_{12}, t_{13}) = \left(A_1^1(X), A_2^1(X), A_3^1(X)\right).$$

This proves (3.59) for $k = 1$. For the other rows, the proof is similar. $\qquad\square$

Remark 3.10. For $t \in \pm O(n)$, we can prove a similar statement.

Chapter 4

Irreducibility: The Case of $m = 3$

Lemma 4.1. *If $\mu^{L_t} \perp \mu$ for all $t \in \mathrm{GL}(3, \mathbb{R}) \setminus \{e\}$, we can approximate by the generators A_{kn}, defined by (3.10), at least one of the following eight triplets of operators:*

$$(x_{1n}, x_{2n}, x_{3n}), (x_{1n}, x_{2n}, D_{3n}), (x_{1n}, D_{2n}, x_{3n}), (D_{1n}, x_{2n}, x_{3n}),$$

$$(x_{1n}, D_{2n}, D_{3n}), (D_{1n}, x_{2n}, D_{3n}), (D_{1n}, D_{2n}, x_{3n}), (D_{1n}, D_{2n}, D_{3n}).$$

Idea of the proof. By Lemma 3.6, the condition of orthogonality $(\mu^3_{(b,a)})^{L_t} \perp \mu^3_{(b,a)}$ for $t \in \pm\mathrm{SL}(3, \mathbb{R}) \setminus \{e\}$ is

$$\Sigma^{\pm}(t) = \Sigma_1^{\pm}(t) + \Sigma_2(t) = \infty, \tag{4.1}$$

where $\Sigma_2(t)$ is defined by (3.34) and $\Sigma_1^{\pm}(t)$ are defined by (3.36). Let $\mathfrak{A}^3$ be the von Neumann algebra generated by the representation.

Cases I and II. We write compactly Lemmas 5.1 and 5.2 as follows, referring to Definition 4.1 for the notation η:

$$x_{rn}x_{rt} \; \eta \; \mathfrak{A}^3 \Leftrightarrow \Delta^{(r)} = \infty, \quad D_{rn} \; \eta \; \mathfrak{A}^3 \Leftrightarrow \Delta_r = \infty, \tag{4.2}$$

$$\text{where} \quad \Delta^{(r)} := \Delta(Y_r^{(r)}, Y_s^{(r)}, Y_t^{(r)}), \quad \Delta_r := \Delta(Y_r, Y_s, Y_t), \tag{4.3}$$

and $\{r, s, t\}$ is a cyclic permutation of $\{1, 2, 3\}$. Here,

$$\|Y_s^{(r)}\|^2 = \sum_{k \in \mathbb{Z}} \frac{b_{rk}^2}{B_{3k}^2 - (b_{1k}^2 + b_{2k}^2 + b_{3k}^2 - b_{sk}^2)}, \quad 1 \leq r, s \leq 3, \qquad (4.4)$$

$$B_{3k} = b_{1k} + b_{2k} + b_{3k}, \quad \text{and} \quad \|Y_r\|^2 = \sum_{k \in \mathbb{Z}} \frac{a_{rk}^2}{\frac{1}{2b_{1k}} + \frac{1}{2b_{2k}} + \frac{1}{2b_{3k}}}. \qquad (4.5)$$

Case III. *Approximation of D_{rn} by $x_{rk}A_{kn}$ and D_{3n} by* $\sin\left(s_k(x_{3k} - a_{3k})\right)A_{kn}$, *(respectively, by* $\cos\left(s_k(x_{3k} - a_{3k})\right)A_{kn}$*). By Lemmas 5.3 and 5.4, we have*

$$D_{rn} \, \eta \, \mathfrak{A}^3 \quad \Leftrightarrow \quad \Delta(Y_{rr}, Y_{rs}, Y_{rt}) = \infty,$$

$$D_{3n} \, \eta \, \mathfrak{A}^3 \quad \Leftrightarrow \quad \Sigma_3(D, s) = \infty, \quad \text{resp.} \quad \Sigma_3^{\vee}(D, s) = \infty,$$

where Y_{kr} for $1 \leq k, r \leq 3$ are defined by (5.13) and $\Sigma_3(D, s)$ and $\Sigma_3^{\vee}(D, s)$ are defined by (5.18). The remainder of this section is devoted to the proof of Lemma 4.1.

4.1   Notations and Change of Variables

In what follows, we systematically use the **notations**:

$$S_r(3) = \sum_{n \in \mathbb{Z}} \frac{b_{rn}^2}{b_{1n}b_{2n} + b_{1n}b_{3n} + b_{2n}b_{3n}}, \quad 1 \leq r \leq 3, \qquad (4.6)$$

$$\Sigma_r := \sum_{n \in \mathbb{Z}} \frac{b_{rn}}{b_{1n} + b_{2n} + b_{3n}}, \quad 1 \leq r \leq 3, \qquad (4.7)$$

$$\Sigma^{rs} := \sum_{k \in \mathbb{Z}} \frac{b_{rk}}{b_{sk}}, \quad 1 \leq r \neq s \leq 3, \quad C_k = \frac{1}{2b_{1k}} + \frac{1}{2b_{2k}} + \frac{1}{2b_{3k}}, \qquad (4.8)$$

$$y_{123} = (y_1, y_2, y_3), \quad \text{where} \quad y_r := \|Y_r\|^2, \qquad (4.9)$$

$$y^{(k)} = (y_1^{(k)}, y_2^{(k)}, y_3^{(k)}) := (\|Y_1^{(k)}\|^2, \|Y_2^{(k)}\|^2, \|Y_3^{(k)}\|^2), \quad 1 \leq k \leq 3, \qquad (4.10)$$

$$y = \begin{pmatrix} y^{(1)} \\ y^{(2)} \\ y^{(3)} \end{pmatrix} = \begin{pmatrix} y_1^{(1)} & y_2^{(1)} & y_3^{(1)} \\ y_1^{(2)} & y_2^{(2)} & y_3^{(2)} \\ y_1^{(3)} & y_2^{(3)} & y_3^{(3)} \end{pmatrix}, \quad \text{where} \quad y_s^{(r)} := \|Y_s^{(r)}\|^2,$$

(4.11)

$$\Sigma_{123}(s) = (\Sigma_{12}(s_{12}), \Sigma_{23}(s_{23}), \Sigma_{13}(s_{13})), \quad s = (s_{12}, s_{23}, s_{13}).$$

(4.12)

We show that at least one of $S_r(3)$ is infinite.

Lemma 4.2. *We have*

$$S_1(3) + S_2(3) + S_3(3) = \infty,$$

(4.13)

$$\|Y_r^{(r)}\|^2 \sim S_r(3) \quad for \ all \quad 1 \le r \le 3,$$

(4.14)

$$\|Y_r^{(s)}\|^2 < \frac{1}{2} S_r(3) \quad for \ all \quad 1 \le r \ne s \le 3,$$

(4.15)

$$\|Y_1^{(i_1)}\|^2 + \|Y_2^{(i_2)}\|^2 + \|Y_3^{(i_3)}\|^2 = \infty, \quad i_1, i_2, i_3 \in \{1, 2, 3\}.$$

(4.16)

Proof. Since $3(a^2 + b^2 + c^2) \ge 2(ab + ac + bc)$, we get

$$S_1(3) + S_2(3) + S_3(3) = \sum_{n \in \mathbb{Z}} \frac{b_{1n}^2 + b_{2n}^2 + b_{3n}^2}{b_{1n}b_{2n} + b_{1n}b_{3n} + b_{2n}b_{3n}} \ge \sum_{k \in \mathbb{Z}} 2/3 = \infty.$$

Further, by (4.4),

$$\|Y_r^{(r)}\|^2 = \sum_{n \in \mathbb{Z}} \frac{b_{rn}^2}{b_{rn}^2 + 2(b_{1n}b_{2n} + b_{1n}b_{3n} + b_{2n}b_{3n})} \stackrel{(3.19)}{\sim} S_r(3),$$

$$\|Y_r^{(s)}\|^2 = \sum_{n \in \mathbb{Z}} \frac{b_{rn}^2}{b_{sn}^2 + 2(b_{1n}b_{2n} + b_{1n}b_{3n} + b_{2n}b_{3n})} < \frac{1}{2} S_r(3), \quad s \ne r.$$

To prove (4.16), we get from (4.4),

$$\|Y_1^{(i_1)}\|^2 + \|Y_2^{(i_2)}\|^2 + \|Y_3^{(i_3)}\|^2$$

$$= \sum_{r=1}^{3} \sum_{n \in \mathbb{Z}} \frac{b_{rn}^2}{b_{i_r n}^2 + 2(b_{1n}b_{2n} + b_{1n}b_{3n} + b_{2n}b_{3n})} > \sum_{n \in \mathbb{Z}} \frac{\sum_{r=1}^{3} b_{rn}^2}{(\sum_{r=1}^{3} b_{rn})^2} = \infty.$$

$\square$

We make the following change of variables:

$$\begin{pmatrix} b_{1n} & b_{2n} & b_{3n} \\ a_{1n} & a_{2n} & a_{3n} \end{pmatrix} \rightarrow \begin{pmatrix} b'_{1n} & b'_{2n} & b'_{3n} \\ a'_{1n} & a'_{2n} & a'_{3n} \end{pmatrix} = \begin{pmatrix} 1 & d_{2n} := \frac{b_{2n}}{b_{1n}} & d_{3n} := \frac{b_{3n}}{b_{1n}} \\ a_{1n}\sqrt{b_{1n}} & a_{2n}\sqrt{b_{1n}} & a_{3n}\sqrt{b_{1n}} \end{pmatrix},$$

$$(4.17)$$

motivated by the following formulas:

$$d\mu_{(b,a)}(x) = \sqrt{\frac{b}{\pi}}\exp(-b(x-a)^2)dx$$

$$= \sqrt{\frac{1}{\pi}}\exp(-(x'-a')^2)dx' = d\mu_{(b',a')}(x'),$$

$$d\mu_{(b_2,a_2)}(x) = \sqrt{\frac{b_2}{\pi}}\exp(-b_2(x-a_2)^2)dx$$

$$= \sqrt{\frac{b_2}{b_1\pi}}\exp\left(-\frac{b_2}{b_1}(x'-a'_2)^2\right)dx'$$

$$= d\mu_{(b'_2,a'_2)}(x'),$$

where $(b',a') = (1, a\sqrt{b})$, $(b'_2, a'_2) = (b_2/b_1, a_2\sqrt{b_1})$.

Remark 4.1. All the expressions given in the list (3.32), (3.33), (3.34) and (4.1) are invariant under the transformations (4.17):

$$S^L_{kr}(\mu) = \sum_{n\in\mathbb{Z}} \frac{b_{kn}}{2}\left(\frac{1}{2b_{rn}} + a_{rn}^2\right),$$

$$Y_r = \left(a_{rk}\left(\frac{1}{2b_{1k}} + \frac{1}{2b_{2k}} + \frac{1}{2b_{3k}}\right)^{-1/2}\right)_{k\in\mathbb{Z}},$$

etc., and $S_r(3)$, which are defined by (4.6).

4.2 Approximation Scheme

Remark 4.2. In what follows, given some expression $< \infty$ (resp., $= \infty$), we denote this case by 0 (resp., by 1).

We use the following notation $S := (S_1(3), S_2(3), S_3(3))$. By Lemma 4.2, we get $\sum_{r=1}^{3} S_r(3) = \infty$. Therefore, without loss of generality, it suffices to consider the following three cases:

$$(1)\ S = (0,0,1), \quad (2)\ S = (0,1,1), \quad (3)\ S = (1,1,1). \qquad (4.18)$$

By Lemma 3.6, the condition of orthogonality $(\mu_{(b,a)}^3)^{L_t} \perp \mu_{(b,a)}^3$ for $t \in \pm \mathrm{SL}(3,\mathbb{R}) \setminus \{e\}$, i.e., $\Sigma^{\pm}(t) = \Sigma_1^{\pm}(t) + \Sigma_2(t) = \infty$, splits into two cases:

$$
\begin{aligned}
&(A)\ \Sigma_1^{\pm}(t) = \infty, \quad \Sigma_1^{\pm}(t) = \sum_{1 \leq i < j \leq 3} \Sigma_{ij}^{\pm}(t), \\
&(B)\ \Sigma_1^{\pm}(t) < \infty \quad \text{but} \quad \Sigma_2(t) = \infty,
\end{aligned}
\qquad (4.19)
$$

where $\Sigma_1^{\pm}(t)$, $\Sigma_{ij}^{\pm}(t)$ and $\Sigma_2(t)$ are defined by (3.36), (3.37) and (3.34).

4.3   Case of $S = (0, 0, 1)$

Lemma 4.3. *The case of $S = (0, 0, 1)$ is equivalent to*

$$\Sigma^{13} + \Sigma^{23} < \infty, \quad S_3(3) \sim \sum_n \frac{b_{3n}^2}{b_{1n} b_{2n}} = \infty. \qquad (4.20)$$

Proof. To prove the first part of (4.20), we set $c_n = \dfrac{b_{3n}}{b_{1n} + b_{2n}}$ and note that

$$
\begin{aligned}
\infty > S_1(3) + S_2(3) \ &= \ \sum_{n \in \mathbb{Z}} \frac{b_{1n}^2 + b_{2n}^2}{b_{1n} b_{2n} + b_{1n} b_{3n} + b_{2n} b_{3n}} \\
&\overset{(3.19)}{\sim} \ \sum_{n \in \mathbb{Z}} \frac{b_{1n}^2 + b_{2n}^2}{(b_{1n} + b_{2n} + b_{3n})^2 - b_{3n}^2} \\
&\sim \ \sum_{n \in \mathbb{Z}} \frac{(b_{1n} + b_{2n})^2}{(b_{1n} + b_{2n} + b_{3n})^2 - b_{3n}^2} \\
&= \ \sum_{n \in \mathbb{Z}} \frac{1}{(1 + c_n)^2 - c_n^2}
\end{aligned}
$$

$$= \sum_{n\in\mathbb{Z}} \frac{1}{1+2c_n} \stackrel{(3.19)}{\sim} \sum_{n\in\mathbb{Z}} \frac{1}{c_n}$$

$$= \sum_{n\in\mathbb{Z}} \frac{b_{1n}+b_{2n}}{b_{3n}} = \Sigma^{13} + \Sigma^{23}.$$

To prove the second part of (4.20), we have from the first part of (4.20),

$$S_3(3) = \sum_{n\in\mathbb{Z}} \frac{b_{3n}^2}{b_{1n}b_{2n} + b_{1n}b_{3n} + b_{2n}b_{3n}}$$

$$= \sum_{n\in\mathbb{Z}} \frac{1}{\frac{b_{1n}b_{2n}}{b_{3n}^2} + \frac{b_{1n}}{b_{3n}} + \frac{b_{2n}}{b_{3n}}} \sim \sum_{n\in\mathbb{Z}} \frac{b_{3n}^2}{b_{1n}b_{2n}}. \qquad \square$$

Lemma 4.4 (Kosyak, 2019). *For any $k \in \mathbb{Z}$, we have*

$$x_{1k}\mathbf{1} \in \langle x_{1k}x_{1n}\mathbf{1} \mid n \in \mathbb{Z} \rangle \Leftrightarrow S_{11}^L(\mu) = \infty.$$

Definition 4.1. A not necessarily bounded self-adjoint operator A in a Hilbert space H is said to be *affiliated* with a von Neumann algebra M of operators in this Hilbert space H if $e^{itA} \in M$ for all $t \in \mathbb{R}$. This is denoted by $A \, \eta \, M$; see Dixmier (1969).

In the case of $S = (0, 0, 1)$, we have

$$\Delta(Y_3^{(3)}, Y_1^{(3)}, Y_2^{(3)}) \sim \Delta(Y_3^{(3)}) \sim \|Y_3^{(3)}\|^2 = \infty,$$

so we can approximate $x_{3n}x_{3t}$ using Lemma 5.1, following which we can approximate x_{3n} using an analogue of Lemma 4.4. *Hereinafter, we consider that x_{3n} can be approximated using Lemma 5.1, without mentioning Lemma 4.4.* We cannot approximate x_{1n} and x_{2n} using Lemma 5.1 since we have

$$\Delta(Y_1^{(1)}, Y_2^{(1)}, Y_3^{(1)}) + \Delta(Y_2^{(2)}, Y_3^{(2)}, Y_1^{(2)}) < \infty.$$

We can try to approximate some of D_{rn} for $1 \leq r \leq 3$ using Lemma 5.2; see Section 4.4.4 for details. We have for $1 \leq k \leq 3$ (see (4.3)),

$$D_{kn} \, \eta \, \mathfrak{A}^3 \Leftrightarrow \Delta_k = \infty, \quad \text{where } \Delta_k := \Delta(Y_k, Y_r, Y_s),$$

and $\{k, r, s\}$ is a cyclic permutation of $\{1, 2, 3\}$. Recall that by $\Sigma^{12} + \Sigma^{13} < \infty$, we get (see (4.5) for the expressions of $\|Y_r\|^2$, $1 \le r \le 3$),

$$\|Y_1\|^2 \sim \sum_{n \in \mathbb{Z}} b_{1n} a_{1n}^2, \quad \|Y_2\|^2 \sim \sum_{n \in \mathbb{Z}} b_{1n} a_{2n}^2, \quad \|Y_3\|^2 \sim \sum_{n \in \mathbb{Z}} b_{1n} a_{3n}^2.$$

(4.21)

By (4.20), we have $\Sigma^{13} + \Sigma^{23} < \infty$. We distinguish two cases:

(1) $\quad \Sigma^{12} < \infty$, and (2) $\quad \Sigma^{12} = \infty$.

In Case (1), since $\Sigma^{12} + \Sigma^{13} < \infty$, we have

$$\infty = S_{1,23}^L(\mu, t, s)$$

$$\overset{(3.47)}{=} \sum_{n \in \mathbb{Z}} \left[\frac{t^2}{4} \frac{b_{1n}}{b_{2n}} + \frac{s^2}{4} \frac{b_{1n}}{b_{3n}} + \frac{b_{1n}}{2} \left(-2a_{1n} + ta_{2n} + sa_{3n} \right)^2 \right]$$

$$\sim \sum_{n \in \mathbb{Z}} \frac{b_{1n}}{2} \left(-2a_{1n} + ta_{2n} + sa_{3n} \right)^2 \overset{(4.21)}{\sim} \|C_1 Y_1 + C_2 Y_2 + C_3 Y_3\|^2.$$

Finally, in Case (1), we can approximate all D_{rn}, $1 \le r \le 3$ using Lemmas 5.2 and 6.4, and the proof is completed. Case (2) can be divided into three cases; if necessary, we can choose an appropriate subsequence of $\left(\frac{b_{1n}}{b_{2n}} \right)_n$:

$$\lim_n \frac{b_{1n}}{b_{2n}} = \begin{cases} (a)\ 0, \\ (b)\ b > 0, \\ (c)\ \infty. \end{cases}$$

(4.22)

Case (c) is reduced to Case (a) by exchanging (b_{2n}, a_{2n}) with (b_{1n}, a_{1n}).

This transformation does not change the first condition in (4.20). In Case (2(a–b)), by (4.5), we obtain the following expressions for $\|Y_r\|^2$, $1 \le r \le 3$:

$$\|Y_1\|^2 = \sum_{n \in \mathbb{Z}} \frac{a_{1n}^2}{\frac{1}{2b_{1n}} + \frac{1}{2b_{2n}} + \frac{1}{2b_{3n}}} = \sum_{k \in \mathbb{Z}} \frac{2b_{1n} a_{1n}^2}{1 + \frac{b_{1n}}{b_{2n}} + \frac{b_{1n}}{b_{3n}}} \overset{\Sigma^{13} < \infty}{\sim} \sum_{n \in \mathbb{Z}} b_{1n} a_{1n}^2,$$

$$\|Y_2\|^2 \sim \sum_{n \in \mathbb{Z}} b_{1n} a_{2n}^2, \quad \|Y_3\|^2 = \sum_{n \in \mathbb{Z}} b_{1n} a_{3n}^2.$$

Since $\|Y_1\|^2 \sim \sum_{n\in\mathbb{Z}} b_{1n} a_{1n}^2 \sim S_{11}^L(\mu) = \infty$, we have four possibilities for $y_{23} := (y_2, y_3) \in \{0,1\}^2$, as in (4.52); see Section 4.4.4:

$$
\begin{array}{ccccc}
 & (1.0) & (1.1) & (1.2) & (1.3) \\
y_1 & 1 & 1 & 1 & 1 \\
y_2 & 0 & 1 & 0 & 1 \\
y_3 & 0 & 0 & 1 & 1
\end{array}\ .
$$

We simply follow the instructions *given in Remark* 4.5. We note that Cases (1.0) and (1.1) cannot occur since the following conditions are contradictory:

$$
S_{13}^L(\mu) \overset{(3.46)}{=} \sum_{n\in\mathbb{Z}} \frac{b_{1n}}{2} \left(\frac{1}{2b_{3n}} + a_{3n}^2 \right) = \infty,
$$

$$
\|Y_3\|^2 \sim \sum_{n\in\mathbb{Z}} b_{1n} a_{3n}^2 < \infty, \quad \Sigma^{13} \overset{(4.20)}{<} \infty.
$$

We have two cases, (1.2.1) and (1.3.1), according to whether the corresponding expressions in (4.57) or (4.58) are divergent. We can approximate in these cases, respectively, D_{1n} and D_{3n}; see (4.54), and all of D_{1n}, D_{2n}, D_{3n}; see (4.55). The proof of irreducibility is finished in both cases because we have $x_{3n}, D_{3n} \ \eta \ \mathfrak{A}^3$ and the problem is reduced to the case of $m = 2$ (Kosyak, 2019) since $A_{kn} = \sum_{r=1}^3 x_{rk} D_{rn} - x_{3k} D_{3n} = \sum_{r=1}^2 x_{rk} D_{rn}$.

If the opposite holds, we have two different cases: (1.2.0) and (1.3.0). We try to approximate D_{3n} using Lemma 5.4. If one of the expressions, $\Sigma_3(D, s)$ or $\Sigma_3^\vee(D, s)$, is divergent for some sequence $s = (s_k)_{k\in\mathbb{Z}}$, we can approximate D_{3k}, and the proof is completed since we have $x_{3n}, D_{3n} \ \eta \ \mathfrak{A}^3$, and the problem is reduced to the case of $m = 2$. Let us suppose, as in Remark 4.7, that for every sequence $s = (s_k)_{k\in\mathbb{Z}}$, we have

$$
\Sigma_3(D, s) + \Sigma_3^\vee(D, s) < \infty.
$$

Then, in particular, we have for $s^{(3)} = (s_k)_{k\in\mathbb{Z}}$, with $\frac{s_k^2}{b_{3k}} \equiv 1$,

$$
\begin{aligned}
\infty \;>\; & \Sigma_3(D, s^{(3)}) + \Sigma_3^\vee(D, s^{(3)}) \sim \Sigma_3(D) + \Sigma_3^\vee(D) \\
=\; & \sum_k \frac{\frac{1}{2b_{3k}} + a_{3k}^2}{C_k + a_{1k}^2 + a_{2k}^2 + a_{3k}^2} \overset{(3.19)}{\sim} \sum_k \frac{\frac{1}{2b_{3k}} + a_{3k}^2}{\frac{1}{2b_{1k}} + a_{1k}^2 + \frac{1}{2b_{2k}} + a_{2k}^2} \\
=\; & \sum_k \frac{\frac{b_{1k}}{b_{3k}} + 2b_{1k}a_{3k}^2}{1 + 2b_{1k}a_{1k}^2 + \frac{b_{1k}}{b_{2k}} + 2b_{1k}a_{2k}^2} \\
\overset{(4.22)}{\sim}\; & \sum_k \frac{2b_{1k}a_{3k}^2}{1 + 2b_{1k}a_{1k}^2 + 2b_{1k}a_{2k}^2} =: \Sigma_3^+(D).
\end{aligned}
$$

Remark 4.3. Finally, we have $\Sigma_3^+(D) \sim \sum_k \frac{2a_{3k}^2}{1 + 2a_{1k}^2 + 2a_{2k}^2}$ since we take $b_{1n} \equiv 1$ by (4.17). In Case (1.2.0), we have $\|Y_2\|^2 \sim \sum_{n\in\mathbb{Z}} b_{1n}a_{2n}^2 < \infty$; therefore, $\Sigma_3^+(D) \sim \sum_k \frac{2a_{3k}^2}{1 + 2a_{1k}^2}$, and hence $\Sigma_3^+(D) = \infty$ by Lemma 4.11. In Case (1.3.0), we have $a_3 = \pm a_1 \pm a_2 + h$ or $a_3 - h = \pm a_1 \pm a_2$; see the proof of Lemma 4.12. Therefore,

$$
\begin{aligned}
\infty > \Sigma_3^+(D) \sim \sum_k \frac{a_{3k}^2}{1 + a_{1k}^2 + a_{2k}^2} &\geq \sum_k \frac{a_{3k}^2}{1 + a_{1k}^2 + 2|a_{1k}||a_{2k}| + a_{2k}^2} \\
= \sum_k \frac{a_{3k}^2}{1 + \left(|a_{1k}| + |a_{2k}|\right)^2}, & \hspace{3em} (4.23)
\end{aligned}
$$

$$
\begin{aligned}
\infty > \Sigma_3^+(D) \sim \sum_k \frac{a_{3k}^2}{1 + a_{1k}^2 + a_{2k}^2} &\geq \sum_k \frac{a_{3k}^2}{1 + a_{1k}^2 + a_{2k}^2 + (|a_{1k}| - |a_{2k}|)^2} \\
\sim \sum_k \frac{a_{3k}^2}{1 + 2a_{1k}^2 - 2|a_{1k}||a_{2k}| + 2a_{2k}^2} & \\
\sim \sum_k \frac{a_{3k}^2}{1 + a_{1k}^2 - 2|a_{1k}||a_{2k}| + a_{2k}^2} \sim \sum_k \frac{a_{3k}^2}{1 + \left(|a_{1k}| - |a_{2k}|\right)^2}. & \quad (4.24)
\end{aligned}
$$

Hence, we have by (4.23) and (4.24),

$$\infty > \Sigma_3^+(D) \geq \sum_k \frac{a_{3k}^2}{1 + \left(\pm a_{1k} \pm a_{2k} \right)^2} = \sum_k \frac{a_{3k}^2}{1 + \left(a_{3k} - h_k \right)^2} = \infty \tag{4.25}$$

by Lemma 4.11; this is a contradiction. Therefore, in both cases, we can approximate D_{3n}, and the proof is completed.

4.4   Case of $S = (0, 1, 1)$

Lemma 4.5. *In the case of $S = (0, 1, 1)$, we have*

$$\lim_n d_{2n} = \lim_n d_{3n} = \infty. \tag{4.26}$$

Proof.   Setting as before $d_{rn} = b_{rn}/b_{1n}$, we obtain by (4.6) and (3.19),

$$S_1(3) = \sum_{n \in \mathbb{Z}} \frac{1}{d_{2n} + d_{3n} + d_{2n}d_{3n}} \overset{(3.19)}{\sim} \sum_{n \in \mathbb{Z}} \frac{1}{(1 + d_{2n})(1 + d_{3n})} < \infty, \tag{4.27}$$

$$S_2(3) = \sum_{n \in \mathbb{Z}} \frac{d_{2n}^2}{d_{2n} + d_{3n} + d_{2n}d_{3n}} \overset{(3.19)}{\sim} \sum_{n \in \mathbb{Z}} \frac{d_{2n}^2}{(1 + d_{2n})(d_{2n} + d_{3n})} = \infty, \tag{4.28}$$

$$S_3(3) = \sum_{n \in \mathbb{Z}} \frac{d_{3n}^2}{d_{2n} + d_{3n} + d_{2n}d_{3n}} \overset{(3.19)}{\sim} \sum_{n \in \mathbb{Z}} \frac{d_{3n}^2}{(1 + d_{3n})(d_{2n} + d_{3n})} = \infty. \tag{4.29}$$

Suppose that $d_{2n} \leq C$ for all $n \in \mathbb{Z}$. Then, by (4.27) and (4.28), we conclude

$$S_1(3) \sim \sum_{n \in \mathbb{Z}} \frac{1}{(1 + d_{2n})(1 + d_{3n})}$$

$$\sim \sum_{n \in \mathbb{Z}} \frac{1}{1 + d_{3n}} \sim \sum_{n \in \mathbb{Z}} \frac{1}{d_{3n}} < \infty,$$

$$\infty = S_2(3) \sim \sum_{n \in \mathbb{Z}} \frac{d_{2n}^2}{(1 + d_{2n})(d_{2n} + d_{3n})} \sim \sum_{n \in \mathbb{Z}} \frac{d_{2n}^2}{d_{2n} + d_{3n}}$$

$$\leq \sum_{n \in \mathbb{Z}} \frac{C^2}{C + d_{3n}} \overset{(3.19)}{\sim} \sum_{n \in \mathbb{Z}} \frac{1}{d_{3n}} < \infty,$$

which is a contradiction. We use the fact that for any fixed $D > 0$, the function $f_D(x) = \frac{x^2}{x+D}$ is strictly increasing when $x > 0$. Similarly, if we suppose that $d_{3n} \leq C$ for all $n \in \mathbb{Z}$, we again obtain a contradiction. $\square$

Lemma 4.6. *The case of $S = (0, 1, 1)$ is equivalent to*

$$S_1(3) \sim \sum_n \frac{b_{1n}^2}{b_{2n} b_{3n}} < \infty, \quad S_2(3) \sim \sum_n \frac{1}{d_n} = \infty, \quad S_3(3) \sim \sum_n d_n = \infty.$$

$$(4.30)$$

Proof. Recall that $d_n = \frac{d_{3n}}{d_{2n}}$. Denote $D_n := 1 + d_{2n}^{-1} + d_{3n}^{-1}$. By Lemma 4.5, we have

$$1 \leq D_n = 1 + d_{2n}^{-1} + d_{3n}^{-1} \leq C, \quad \text{for all} \quad n \in \mathbb{Z}. \qquad (4.31)$$

Therefore, we get

$$S_1(3) = \sum_{n \in \mathbb{Z}} \frac{1}{d_{2n} + d_{3n} + d_{2n} d_{3n}} = \sum_{n \in \mathbb{Z}} \frac{1}{D_n d_{2n} d_{3n}} \sim \sum_{n \in \mathbb{Z}} \frac{1}{d_{2n} d_{3n}}$$

$$= \sum_n \frac{b_{1n}^2}{b_{2n} b_{3n}},$$

$$S_2(3) = \sum_{n \in \mathbb{Z}} \frac{d_{2n}^2}{d_{2n} + d_{3n} + d_{2n} d_{3n}} = \sum_{n \in \mathbb{Z}} \frac{d_{2n}^2}{D_n d_{2n} d_{3n}} \sim \sum_{n \in \mathbb{Z}} \frac{1}{d_n},$$

$$S_3(3) = \sum_{n \in \mathbb{Z}} \frac{d_{3n}^2}{d_{2n} + d_{3n} + d_{2n} d_{3n}} = \sum_{n \in \mathbb{Z}} \frac{d_{3n}^2}{D_n d_{2n} d_{3n}} \sim \sum_{n \in \mathbb{Z}} d_n. \qquad \square$$

By Lemma 4.2, (4.15), we get $\|Y_1^{(r)}\|^2 < \infty$, $1 \leq r \leq 3$; therefore, we get the following Lemma 4.7.

Lemma 4.7. *In the case of $S = (0, 1, 1)$, we have*

$$\Delta(Y_1^{(1)}, Y_2^{(1)}, Y_3^{(1)}) < \infty, \quad \Delta(Y_2^{(2)}, Y_3^{(2)}, Y_1^{(2)}) \sim \Delta(Y_2^{(2)}, Y_3^{(2)}),$$

$$\Delta(Y_3^{(3)}, Y_1^{(3)}, Y_2^{(3)}) \sim \Delta(Y_3^{(3)}, Y_2^{(3)}). \tag{4.32}$$

Proof. Set $(f_1, f_2, f_3) = (Y_3^{(3)}, Y_1^{(3)}, Y_2^{(3)})$. Then,

$$\Delta(f_1, f_2, f_3) \overset{(3.15)}{=} \frac{\Gamma(f_1) + \Gamma(f_1, f_2) + \Gamma(f_1, f_3) + \Gamma(f_1, f_2, f_3)}{1 + \Gamma(f_2) + \Gamma(f_3) + \Gamma(f_2, f_3)}$$

$$> \frac{\Gamma(f_1) + \Gamma(f_1, f_3)}{1 + \Gamma(f_2) + \Gamma(f_3) + \Gamma(f_2, f_3)}$$

$$\overset{(4.33)}{\geq} \frac{\Gamma(f_1) + \Gamma(f_1, f_3)}{(1 + \Gamma(f_2))(1 + \Gamma(f_3))} \sim \Delta(f_1, f_3)$$

since $f_2 \in l_2(\mathbb{Z})$; see (3.14). Indeed, for $f, g \in l_2(\mathbb{Z})$ and $f \in l_2(\mathbb{Z}), g \notin l_2(\mathbb{Z})$, we have, respectively,

$$\Gamma(f, g) \leq \Gamma(f)\Gamma(g) < \infty, \quad \Gamma(f, g) \leq \Gamma(f)\Gamma(g),$$

where $\Gamma(f, g), \Gamma(g)$ are defined by

$$\Gamma(f, g) := \lim_n \Gamma(f_{(n)}, g_{(n)}), \quad \Gamma(g) := \lim_n \Gamma(g_{(n)}), \tag{4.33}$$

and $g_{(n)} := (g_k)_{k=-n}^n \in \mathbb{R}^{2n+1}$. Similarly, set $(f_1, f_2, f_3) = (Y_2^{(2)}, Y_3^{(2)}, Y_1^{(2)})$. Then,

$$\Delta(f_1, f_2, f_3) \overset{(3.15)}{=} \frac{\Gamma(f_1) + \Gamma(f_1, f_2) + \Gamma(f_1, f_3) + \Gamma(f_1, f_2, f_3)}{1 + \Gamma(f_2) + \Gamma(f_3) + \Gamma(f_2, f_3)}$$

$$> \frac{\Gamma(f_1) + \Gamma(f_1, f_2)}{1 + \Gamma(f_2) + \Gamma(f_3) + \Gamma(f_2, f_3)}$$

$$\overset{(4.33)}{\geq} \frac{\Gamma(f_1) + \Gamma(f_1, f_2)}{(1 + \Gamma(f_2))(1 + \Gamma(f_3))} \sim \Delta(f_1, f_2)$$

since $f_3 \in l_2(\mathbb{Z})$. Finally, we derive both equivalences in (4.32). To prove $\Delta(Y_1^{(1)}, Y_2^{(1)}, Y_3^{(1)}) < \infty$, we set $(f_1, f_2, f_3) = (Y_1^{(1)}, Y_2^{(1)}, Y_3^{(1)})$, and note that

$$\Delta(f_1, f_2, f_3) \overset{(3.15)}{=} \frac{\Gamma(f_1) + \Gamma(f_1, f_2) + \Gamma(f_1, f_3) + \Gamma(f_1, f_2, f_3)}{1 + \Gamma(f_2) + \Gamma(f_3) + \Gamma(f_2, f_3)}$$

$$\leq \frac{\Gamma(f_1)\left(1 + \Gamma(f_2) + \Gamma(f_3) + \Gamma(f_2, f_3)\right)}{1 + \Gamma(f_2) + \Gamma(f_3) + \Gamma(f_2, f_3)} = \Gamma(f_1) < \infty. \qquad \square$$

In order to approximate x_{2n} or x_{3n}, it remains to study the case of

$$\Delta(Y_2^{(2)}, Y_3^{(2)}) = \infty, \quad \Delta(Y_3^{(3)}, Y_2^{(3)}) = \infty, \tag{4.34}$$

where $\Delta(f_1, f_2) = \frac{\Gamma(f_1) + \Gamma(f_1, f_2)}{1 + \Gamma(f_2)}$. For $2 \leq r \leq 3$, denote

$$\rho_r(C_2, C_3) := \|C_2 Y_2^{(r)} + C_3 Y_3^{(r)}\|^2, \quad (C_2, C_3) \in \mathbb{R}^2, \tag{4.35}$$

$$\nu(C_1, C_2, C_3) := \|C_1 Y_1 + C_2 Y_2 + C_3 Y_3\|^2, \quad (C_1, C_2, C_3) \in \mathbb{R}^3. \tag{4.36}$$

Lemma 4.8. *In the case of $S = (0, 1, 1)$, we have*

$$\rho_2(C_2, C_3) \sim \sum_{n \in \mathbb{Z}} \frac{(C_2 + C_3 d_n)^2}{1 + 2d_n},$$

$$\rho_3(C_2, C_3) \sim \sum_{n \in \mathbb{Z}} \frac{(C_2 + C_3 d_n)^2}{d_n^2 + 2d_n} = \sum_{n \in \mathbb{Z}} \frac{(C_2 l_n + C_3)^2}{1 + 2l_n}, \tag{4.37}$$

$$\nu(C_1, C_2, C_3) \sim \sum_{n \in \mathbb{Z}} b_{1n} \left(\sum_{r=1}^{3} C_r a_{rn} \right)^2. \tag{4.38}$$

Proof. Set as before $d_n = \frac{d_{3n}}{d_{2n}}$. By (4.4) and (4.5), we get

$$\|Y_2^{(2)}\|^2 = \sum_{n \in \mathbb{Z}} \frac{d_{2n}^2}{d_{2n}^2 + 2(d_{2n} + d_{3n} + d_{2n}d_{3n})} = \sum_{n \in \mathbb{Z}} \frac{d_{2n}^2}{d_{2n}^2 + 2D_n d_{2n} d_{3n}}$$

$$\sim \sum_{n \in \mathbb{Z}} \frac{1}{D_n d_n},$$

$$\|Y_3^{(2)}\|^2 = \sum_{n\in\mathbb{Z}} \frac{d_{3n}^2}{d_{2n}^2 + 2(d_{2n} + d_{3n} + d_{2n}d_{3n})} = \sum_{n\in\mathbb{Z}} \frac{d_{3n}^2}{d_{2n}^2 + 2D_n d_{2n} d_{3n}}$$

$$= \sum_{n\in\mathbb{Z}} \frac{d_n^2}{1 + 2D_n d_n},$$

$$\|Y_2^{(3)}\|^2 = \sum_{n\in\mathbb{Z}} \frac{d_{2n}^2}{d_{3n}^2 + 2(d_{2n} + d_{3n} + d_{2n}d_{3n})} = \sum_{n\in\mathbb{Z}} \frac{d_{2n}^2}{d_{3n}^2 + 2D_n d_{2n} d_{3n}}$$

$$= \sum_{n\in\mathbb{Z}} \frac{1}{d_n^2 + 2D_n d_n}, \tag{4.39}$$

$$\|Y_3^{(3)}\|^2 = \sum_{n\in\mathbb{Z}} \frac{d_{3n}^2}{d_{3n}^2 + 2(d_{2n} + d_{3n} + d_{2n}d_{3n})}$$

$$= \sum_{n\in\mathbb{Z}} \frac{d_{3n}^2}{d_{3n}^2 + 2D_n d_{2n} d_{3n}} \sim \sum_{n\in\mathbb{Z}} \frac{d_n}{D_n},$$

$$\|Y_1\|^2 = \sum_{n\in\mathbb{Z}} \frac{a_{1n}^2}{\frac{1}{2b_{1n}} + \frac{1}{2b_{2n}} + \frac{1}{2b_{3n}}} = \sum_{k\in\mathbb{Z}} \frac{2b_{1n} a_{1n}^2}{1 + d_{2n}^{-1} + d_{3n}^{-1}} = \sum_{n\in\mathbb{Z}} \frac{2b_{1n} a_{1n}^2}{D_n},$$

$$\|Y_2\|^2 = \sum_{n\in\mathbb{Z}} \frac{2b_{1n} a_{2n}^2}{D_n}, \quad \|Y_3\|^2 = \sum_{n\in\mathbb{Z}} \frac{b_{1n} a_{3n}^2}{D_n}. \tag{4.40}$$

Recall that $d_{rn} = \frac{b_{rn}}{b_{1n}}$. By (4.31), we obtain

$$\|Y_2^{(2)}\|^2 \sim \sum_{n\in\mathbb{Z}} \frac{1}{1 + 2d_n} \sim \sum_{n\in\mathbb{Z}} \frac{1}{d_n}, \quad \|Y_3^{(2)}\|^2 \sim \sum_{n\in\mathbb{Z}} \frac{d_n^2}{1 + 2d_n}, \tag{4.41}$$

$$\|Y_2^{(3)}\|^2 \sim \sum_{n\in\mathbb{Z}} \frac{1}{d_n^2 + 2d_n}, \quad \|Y_3^{(3)}\|^2 \sim \sum_{n\in\mathbb{Z}} \frac{d_n^2}{d_n^2 + 2d_n} \sim \sum_{n\in\mathbb{Z}} d_n,$$

$$\|Y_1\|^2 \sim \sum_{n\in\mathbb{Z}} b_{1n} a_{1n}^2, \quad \|Y_2\|^2 \sim \sum_{n\in\mathbb{Z}} b_{1n} a_{2n}^2, \quad \|Y_3\|^2 \sim \sum_{n\in\mathbb{Z}} b_{1n} a_{3n}^2,$$

$$\|C_1 Y_1 + C_2 Y_2 + C_3 Y_3\|^2 \overset{(4.31)}{\sim} \sum_{n\in\mathbb{Z}} b_{1n}\left(C_1 a_{1n} + C_2 a_{2n} + C_3 a_{3n}\right)^2.$$

$$\tag{4.42}$$

By (4.41) and (4.42), the proof is completed. $\qquad\qquad\square$

4.4.1 *Approximation of x_{2n}, x_{3n}*

To approximate x_{2n}, x_{3n}, we need several lemmas. Denote $l_n = d_n^{-1}$.

Lemma 4.9. *The following five series are equivalent:*

$$\text{(i} - \text{ii)} \quad \sum_{n\in\mathbb{Z}} \frac{(C_2 - C_3 d_n)^2}{1 + 2d_n} \sim \sum_{n\in\mathbb{Z}} c_n^2, \tag{4.43}$$

$$\text{(iii} - \text{iv)} \quad \sum_{n\in\mathbb{Z}} \frac{(C_2 l_n - C_3)^2}{1 + 2l_n} \sim \sum_{n\in\mathbb{Z}} e_n^2, \tag{4.44}$$

$$\text{(v)}\ \Sigma_{23}(s) = \sum_{n\in\mathbb{Z}} \left(s^2 \sqrt{\frac{b_{2n}}{b_{3n}}} - s^{-2} \sqrt{\frac{b_{3n}}{b_{2n}}} \right)^2 = \sum_{n\in\mathbb{Z}} \left(\frac{s^2}{\sqrt{d_n}} - \frac{\sqrt{d_n}}{s^2} \right)^2,$$

$$\tag{4.45}$$

where

$$d_n = C_2 C_3^{-1}(1 + c_n), \quad l_n = C_3 C_2^{-1}(1 + e_n),$$
$$s^4 = C_2 C_3^{-1} > 0, \quad l_n = d_n^{-1}. \tag{4.46}$$

Proof. To prove (4.43) and (4.44), we get by Lemma 3.3 using (4.46)

$$\sum_{n\in\mathbb{Z}} \frac{(C_2 - C_3 d_n)^2}{1 + 2d_n} = \sum_{n\in\mathbb{Z}} \frac{C_2^2 c_n^2}{1 + 2C_2 C_3^{-1}(1 + c_n)} \sim \sum_{n\in\mathbb{Z}} c_n^2,$$

$$\sum_{n\in\mathbb{Z}} \frac{(C_2 l_n - C_3)^2}{1 + 2l_n} = \sum_{n\in\mathbb{Z}} \frac{C_3^2 e_n^2}{1 + 2C_3 C_2^{-1}(1 + e_n)} \sim \sum_{n\in\mathbb{Z}} e_n^2.$$

To complete the proof, we make use of the following lemma. $\square$

Lemma 4.10. *Let $(c_n)_{n\in\mathbb{Z}}$ be a sequence of real numbers, with $1 + c_n > 0$ and $(1 + c_n)(1 + e_n) = 1$. Then, the following three series are equivalent:*

$$\sum_{n\in\mathbb{Z}} \left((1 + e_n)^{1/2} - (1 + e_n)^{-1/2}\right)^2, \quad \sum_{n\in\mathbb{Z}} c_n^2 \quad and \quad \sum_{n\in\mathbb{Z}} e_n^2.$$

Proof. Set $s^4 = C_2 C_3^{-1}$, replacing $1 + c_n$ by $(1 + e_n)^{-1}$ in Lemma 6.8 gives

$$\Sigma_{23}(s) = \sum_{n \in \mathbb{Z}} ((1 + c_n)^{-1/2} - (1 + c_n)^{1/2})^2$$

$$= \sum_{n \in \mathbb{Z}} ((1 + e_n)^{1/2} - (1 + e_n)^{-1/2})^2.$$

Therefore, $\sum_{n \in \mathbb{Z}} \frac{c_n^2}{1+c_n} = \sum_{n \in \mathbb{Z}} \frac{e_n^2}{1+e_n}$, and hence, by Lemma 3.3, the two series are equivalent: $\sum_{n \in \mathbb{Z}} c_n^2 \sim \sum_{n \in \mathbb{Z}} e_n^2$. $\qquad\square$

4.4.2   *Two remaining possibilities*

By Lemma 4.9, there are only two cases:

(1) when $\rho_2(C_2, C_3) = \rho_3(C_2, C_3) = \infty$ for all $(C_2, C_3) \in \mathbb{R}^2 \setminus \{0\}$;
(2) when both $\rho_2(C_2, C_3)$ and $\rho_3(C_2, C_3)$ are finite, and hence $\Sigma_{23}(s) < \infty$.

To illustrate this, we start with the following example.

Example 4.1. Set $d_n = n^\alpha$ for $n \in \mathbb{N}$, with $\alpha \in \mathbb{R}$. We have

$$\lim_n d_n = \begin{cases} \infty, & \text{if } \alpha > 0, \\ 1, & \text{if } \alpha = 0, \\ 0, & \text{if } \alpha < 0. \end{cases} \tag{4.47}$$

For the general sequence $(d_n)_{n \in \mathbb{Z}}$, we have four cases (if necessary, we can chose an appropriate subsequence):

$$\lim_n d_n = \begin{cases} (a) \ \infty, \\ (b) \ d > 0 \quad \text{with } \sum_n c_n^2 = \infty, \\ (c) \ d > 0 \quad \text{with } \sum_n c_n^2 < \infty, \\ (d) \ d = 0, \end{cases} \tag{4.48}$$

where $d_n = d(1 + c_n)$ and $\lim_n c_n = 0$.

4.4.3 Cases (a), (b) and (d)

Remark 4.4. In Case (a), we see by (4.37) that

$$\rho_2(C_2, C_3) = \rho_3(C_2, C_3) = \infty \quad \text{for all } (C_2, C_3) \in \mathbb{R}^2 \setminus \{0\}.$$

Case (d) is reduced to Case (a) by exchanging (b_{2n}, a_{2n}) with (b_{3n}, a_{3n}). In Case (b), by Lemma 4.9 and (4.48), we conclude that

$$\rho_2(C_2, -C_3) = \rho_3(C_2, -C_3) = \infty \quad \text{for } C_2 C_3^{-1} > 0.$$

Hence, $\rho_2(C_2, C_3) = \rho_3(C_2, C_3) = \infty$ for all $(C_2, C_3) \in \mathbb{R}^2 \setminus \{0\}$. Therefore, in Cases (a), (b) and (d), we get $x_{2n}, x_{3n} \, \eta \, \mathfrak{A}^3$.

To complete the proof for these cases, it is sufficient to approximate one of the operators D_{rn}, $1 \leq r \leq 3$ by the operators $(A_{kn})_{k \in \mathbb{Z}}$ using Lemmas 5.2; see Section 4.4.4. Alternatively, we can try to approximate D_{3n}, D_{2n} using Lemma 5.4 and its analogue; see Section 4.4.5.

Note that by Lemma 4.5, we have $\lim_n b_{2n} = \lim_n b_{3n} = \infty$. In Cases (a) and (b), the conditions (4.30) are expressed by (4.48) as follows:

$$b = (1, b_{2n}, d_n b_{2n}), \quad \sum_n \frac{1}{b_{2n}^2 d_n} < \infty, \quad \sum_n \frac{1}{d_n} = \infty, \quad \lim_n d_n = \infty,$$
$$\tag{4.49}$$

$$b = (1, b_{2n}, db_{2n}(1 + c_n)), \quad \sum_n \frac{1}{b_{2n}^2} < \infty, \quad \sum_n c_n^2 = \infty. \tag{4.50}$$

Indeed, to get (4.49), we observe that (4.30) are expressed as follows:

$$S_1(3) \sim \sum_n \frac{1}{b_{2n} b_{3n}} = \sum_n \frac{1}{b_{2n}^2 d_n} < \infty, \quad S_2(3) \sim \sum_n \frac{1}{d_n} = \infty.$$

The condition $S_3(3) \sim \sum_n d_n = \infty$ holds by $\lim_n d_n = \infty$.

In order to get (4.50), we express the conditions (4.30) as follows:

$$S_1(3) \sim \sum_n \frac{1}{b_{2n} db_{2n}(1 + c_n)} \sim \sum_n \frac{1}{b_{2n}^2} < \infty,$$

$$S_2(3) \sim \sum_n \frac{1}{d_n} = \sum_n \frac{1}{1 + c_n} = \infty, \quad S_3(3) \sim \sum_n d_n = \sum_n (1 + c_n) = \infty.$$

The condition $S_2(3) = \infty$ holds by $\lim_n c_n = 0$.

4.4.4   *Approximation of D_{rn}, $1 \leq r \leq 3$*

By Lemma 5.2, we have for $1 \leq k \leq 3$ (see (4.3)),

$$D_{kn} \, \eta \, \mathfrak{A}^3 \Leftrightarrow \Delta_k = \infty, \quad \text{where } \Delta_k := \Delta(Y_k, Y_r, Y_s),$$

and $\{k, r, s\}$ is a cyclic permutation of $\{1, 2, 3\}$. Recall that by (4.40),

$$\|Y_1\|^2 \sim \sum_{n \in \mathbb{Z}} b_{1n} a_{1n}^2, \quad \|Y_2\|^2 \sim \sum_{n \in \mathbb{Z}} b_{1n} a_{2n}^2, \quad \|Y_3\|^2 \sim \sum_{n \in \mathbb{Z}} b_{1n} a_{3n}^2.$$

$$(4.51)$$

Since $\|Y_1\|^2 \sim \sum_{n \in \mathbb{Z}} b_{1n} a_{1n}^2 \sim S_{11}^L(\mu) = \infty$, we have four possibilities for $y_{23} := (y_2, y_3) \in \{0, 1\}^2$:

$$
\begin{array}{c|cccc}
 & (1.0) & (1.1) & (1.2) & (1.3) \\
\hline
y_1 & 1 & 1 & 1 & 1 \\
y_2 & 0 & 1 & 0 & 1 \\
y_3 & 0 & 0 & 1 & 1
\end{array}
\qquad (4.52)
$$

In Case (1.0), we have $\Delta(Y_1, Y_2, Y_3) \sim \|Y_1\|^2 = \infty$; therefore, we can approximate D_{1n} using Lemma 5.3, and the proof is completed. We should consider the following three cases: (1.1), (1.2) and (1.3). In Cases (1.1), (1.2) and (1.3), we have, respectively (see the proof of Lemma 4.7),

$$\Delta(Y_1, Y_2, Y_3) \sim \Delta(Y_1, Y_2), \quad \Delta(Y_2, Y_3, Y_1) \sim \Delta(Y_2, Y_1), \quad (4.53)$$

$$\Delta(Y_1, Y_2, Y_3) \sim \Delta(Y_1, Y_3), \quad \Delta(Y_3, Y_1, Y_2) \sim \Delta(Y_3, Y_1), \quad (4.54)$$

$$\Delta(Y_1, Y_2, Y_3), \quad \Delta(Y_2, Y_3, Y_1), \quad \Delta(Y_3, Y_1, Y_2). \quad (4.55)$$

By (4.40) and Lemma 4.5, we have, respectively, in cases (1.1)–(1.3),

$$\nu_{12}(C_1, C_2) := \|C_1 Y_1 + C_2 Y_2\|^2 \sim \sum_{n \in \mathbb{Z}} b_{1n}(C_1 a_{1n} + C_2 a_{2n})^2, \quad (4.56)$$

$$\nu_{13}(C_1, C_3) := \|C_1 Y_1 + C_3 Y_3\|^2 \sim \sum_{n \in \mathbb{Z}} b_{1n}(C_1 a_{1n} + C_3 a_{3n})^2, \quad (4.57)$$

$$\nu(C_1, C_2, C_3) = \sum_{n \in \mathbb{Z}} b_{1n}(C_1 a_{1n} + C_2 a_{2n} + C_3 a_{3n})^2. \quad (4.58)$$

Remark 4.5. We have three cases, (1.1.1), (1.2.1) and (1.3.1), according to whether the corresponding expressions in (4.56), (4.57) and (4.58) are divergent. We can approximate in these cases D_{1n} and D_{2n} in (4.53), D_{1n} and D_{3n} in (4.54) and D_{1n}, D_{2n}, D_{3n} in (4.55). The proof of irreducibility is thus completed in these cases because we have $D_{rn}, x_{2n}, x_{3n} \eta \, \mathfrak{A}^3$ for some $1 \leq r \leq 3$. If the opposite holds, we have three different cases:

$$(1.1.0) \quad \|C_1 Y_1 + C_2 Y_2\| < \infty \quad \text{for some } (C_1, C_2) \in \mathbb{R}^2 \setminus \{0\},$$

$$(1.2.0) \quad \|C_1 Y_1 + C_3 Y_3\| < \infty \quad \text{for some } (C_1, C_3) \in \mathbb{R}^2 \setminus \{0\},$$

$$(1.3.0) \quad \nu(C_1, C_2, C_3) < \infty \quad \text{for some } (C_1, C_2, C_3) \in \mathbb{R}^3 \setminus \{0\}.$$

Recall that by (3.47), we have

$$S_{1,23}^L(\mu, t, s) = \sum_{n \in \mathbb{Z}} \left[\frac{t^2}{4} \frac{b_{1n}}{b_{2n}} + \frac{s^2}{4} \frac{b_{1n}}{b_{3n}} + \frac{b_{1n}}{2}(-2a_{1n} + ta_{2n} + sa_{3n})^2 \right].$$

Remark 4.6. In Case (1.1.0), we have $\Sigma^{12} = \sum_{n \in \mathbb{Z}} \frac{b_{1n}}{b_{2n}} = \infty$ since

$$S_{1,23}^L(\mu, t, 0) = \infty, \quad \text{but } \nu_{12}(C_1, C_2) < \infty.$$

Similarly, $\Sigma^{13} = \infty$ since $S_{13}^L(\mu) = \sum_{n \in \mathbb{Z}} \frac{b_{1n}}{2}\left(\frac{1}{2b_{3n}} + a_{3n}^2\right) = \infty$, but $\|Y_3\|^2 \sim \sum_{n \in \mathbb{Z}} b_{1n}a_{3n}^2 < \infty$; see (3.46) for the definition of $S_{kr}^L(\mu)$.

In Case (1.2.0), we conclude that $\Sigma^{13} = \infty$ since $S_{1,23}^L(\mu, 0, s) = \infty$, but $\nu_{13}(C_1, C_3) < \infty$. Similarly, $\Sigma^{12} = \infty$ since

$$S_{12}^L(\mu) = \sum_{n \in \mathbb{Z}} \frac{b_{1n}}{2}\left(\frac{1}{2b_{2n}} + a_{2n}^2\right) = \infty, \quad \text{but } \|Y_2\|^2 \sim \sum_{n \in \mathbb{Z}} b_{1n}a_{2n}^2 < \infty.$$

In Case (1.3.0), we have $\Sigma^{12} = \Sigma^{13} = \infty$ since $S_{1,23}^L(\mu, t, s) = \infty$, but $\nu(C_1, C_2, C_3) \sim \sum_{n \in \mathbb{Z}} b_{1n}(C_1 a_{1n} + C_2 a_{2n} + C_3 a_{3n})^2 < \infty$.

So, it remains to consider only the three following cases, when $\Sigma^{12} = \Sigma^{13} = \infty$:

$$(1.1.0), \quad (1.2.0), \quad (1.3.0).$$

By Lemma 5.3, we have

$$D_{2n} \, \eta \, \mathfrak{A}^3 \Leftrightarrow \Delta(Y_{22}, Y_{23}, Y_{21}) = \infty, \quad D_{3n} \, \eta \, \mathfrak{A}^3 \Leftrightarrow \Delta(Y_{33}, Y_{31}, Y_{32}) = \infty,$$

where vectors Y_{rs} for $2 \leq r \leq 3$, $1 \leq s \leq 3$ are defined by (5.12)–(5.13). We cannot prove that $\Delta(Y_{22}, Y_{23}, Y_{21}) = \infty$ or $\Delta(Y_{33}, Y_{31}, Y_{32}) = \infty$. Therefore, to approximate D_{3n}, we are forced to prove Lemma 5.4 and its analogue for D_{2n}; see Remark 4.7 later.

4.4.5   *Two technical lemmas*

Lemma 4.11. *Let $a_1, a_2 \notin l_2(\mathbb{Z})$ and $C_1 a_1 + C_2 a_2 \in l_2(\mathbb{Z})$ for some $(C_1, C_2) \in \mathbb{R}^2 \setminus \{0\}$, where $a_r = (a_{rk})_{k \in \mathbb{Z}}$, $1 \leq r \leq 2$. Then, we have*

$$\sum_{k \in \mathbb{Z}} \frac{a_{1k}^2}{1 + a_{2k}^2} = \infty. \tag{4.59}$$

Proof. We set $Y_r = a_r$, in Case (1.1.0) when $C_1 Y_1 + C_2 Y_2 = h \in l_2(\mathbb{Z})$ with $C_1 C_2 > 0$ (where we have $C_1 C_2 \neq 0$), we should take $a_2 = -a_1 + h$, and in the case when $C_1 C_2 < 0$, we take $a_2 = a_1 + h$. The series $\sum_{k \in \mathbb{Z}} \frac{a_{1k}^2}{1 + a_{2k}^2}$ will remain equivalent to the initial one if we replace (C_1, C_2) with $(\pm 1, 1)$ in the expression for h. Fix a small $\varepsilon > 0$ and a large $N \in \mathbb{N}$. Since $|\pm a + b| \leq |a| + |b|$, we get

$$\sum_{k \in \mathbb{Z}} \frac{a_{1k}^2}{1 + a_{2k}^2} = \sum_{k \in \mathbb{Z}} \frac{a_{1k}^2}{1 + (\pm a_{1k} + h_k)^2} \geq \sum_{k \in \mathbb{Z}} \frac{a_{1k}^2}{1 + a_{1k}^2 + 2|a_{1k}||h_k| + h_k^2}$$

$$\overset{(3.19)}{\sim} \sum_{k \in \mathbb{Z}} \frac{a_{1k}^2}{1 + 2|a_{1k}||h_k| + h_k^2} \overset{(*)}{>} \sum_{k \in \mathbb{Z}_N} \frac{a_{1k}^2}{1 + 2|a_{1k}|\varepsilon + \varepsilon^2}$$

$$\overset{(3.17)}{\sim} \sum_{k \in \mathbb{Z}_N} a_{1k}^2 = \infty,$$

where $\mathbb{Z}_N := \{n \in \mathbb{Z} \mid |n| > N\}$. The inequality $(*)$ holds since $h \in l_2(\mathbb{Z})$, and we have $\sum_{k \in \mathbb{Z}_N} h_k^2 < \varepsilon^2$ for sufficiently large $N \in \mathbb{N}$. $\qquad\square$

Lemma 4.12. *Let $a_1, a_2, a_3 \notin l_2(\mathbb{Z})$ and $C_1 a_1 + C_2 a_2 + C_3 a_3 \in l_2(\mathbb{Z})$ for some $(C_1, C_2, C_3) \in \mathbb{R}^3$, $C_3 \neq 0$, where $a_r = (a_{rk})_{k \in \mathbb{Z}}$ for $1 \leq r \leq 3$. Then, we have*

$$\sum_{k \in \mathbb{Z}} \frac{a_{1k}^2 + a_{2k}^2}{1 + a_{3k}^2} = \infty. \tag{4.60}$$

Proof. We set $Y_r = a_r$, in Case (1.3.0), and we have $C_1 a_1 + C_2 a_2 + C_3 a_3 = h \in l_2(\mathbb{Z})$ for some $(C_1, C_2, C_3) \in \mathbb{R}^3$; see Remark 4.5. We can take $C_3 = 1$. Then, $a_3 = -C_1 a_1 - C_2 a_2 + h$. When $C_1 = 0$ or $C_2 = 0$, the lemma is reduced to Lemma 4.11. Suppose $C_1 C_2 \neq 0$. The series $\sum_{k \in \mathbb{Z}} \frac{a_{1k}^2 + a_{2k}^2}{1 + a_{3k}^2}$ will remain equivalent to the initial one if we replace (C_1, C_2, C_3) with $(\pm 1, \pm 1, 1)$ in the expression for h. Fix a small $\varepsilon > 0$ and a large $N \in \mathbb{N}$. Suppose the opposite, i.e.,

$$\infty > \sum_{k \in \mathbb{Z}} \frac{a_{1k}^2 + a_{2k}^2}{1 + \left(\pm a_{1k} \pm a_{2k} + h_k \right)^2}, \quad \text{then}$$

$$\infty > \sum_{k \in \mathbb{Z}} \frac{(|a_{1k}| + |a_{2k}|)^2}{1 + \left(\pm a_{1k} \pm a_{2k} + h_k \right)^2}$$

$$\geq \sum_{k \in \mathbb{Z}} \frac{(|a_{1k}| + |a_{2k}|)^2}{1 + a_{1k}^2 + a_{2k}^2 + 2|a_{1k}||a_{2k}| + 2|a_{1k}||h_k| + 2|a_{2k}||h_k| + h_k^2}$$

$$\overset{(3.19)}{\sim} \sum_{k \in \mathbb{Z}} \frac{(|a_{1k}| + |a_{2k}|)^2}{1 + 2|a_{1k}||h_k| + 2|a_{2k}||h_k| + h_k^2}$$

$$\overset{(*)}{>} \sum_{k \in \mathbb{Z}_N} \frac{(|a_{1k}| + |a_{2k}|)^2}{1 + 2(|a_{1k}| + |a_{2k}|)\varepsilon + \varepsilon^2}$$

$$\overset{(3.17)}{\sim} \sum_{k \in \mathbb{Z}_N} (|a_{1k}| + |a_{2k}|)^2 = \infty,$$

where $\mathbb{Z}_N := \{n \in \mathbb{Z} \mid |n| > N\}$, which is a contradiction. The inequality $(*)$ holds since $h \in l_2(\mathbb{Z})$, and we have $\sum_{k \in \mathbb{Z}_N} h_k^2 < \varepsilon^2$ for sufficiently large $N \in \mathbb{N}$. $\qquad \square$

Remark 4.7. It is possible to prove an analogue of Lemma 5.4 to approximate D_{2n} with the corresponding expressions $\Sigma_2(D,s)$, $\Sigma_2^\vee(D,s)$ and $\Sigma_3(D)$, $\Sigma_3^\vee(D)$. If one of the expressions $\Sigma_2(D,s)$, $\Sigma_2^\vee(D,s)$, $\Sigma_3(D,s)$ or $\Sigma_3^\vee(D,s)$ is divergent for some sequence $s = (s_k)_{k\in\mathbb{Z}}$, we can approximate D_{2k} or D_{3k}, and the proof is completed when $S = (0,1,1)$ in Cases (a) and (b). Suppose that for all sequence $s = (s_k)_{k\in\mathbb{Z}}$, we have

$$\Sigma_2(D,s) + \Sigma_2^\vee(D,s) + \Sigma_3(D,s) + \Sigma_3^\vee(D,s) < \infty.$$

Then, in particular, we have for $s^{(r)} = (s_{rk})_{k\in\mathbb{Z}}$, $2 \le r \le 3$ with $\frac{s_{rk}^2}{b_{rk}} \equiv 1$,

$$\begin{aligned}
\infty \;>\; & \Sigma_2(D,s^{(2)}) + \Sigma_2^\vee(D,s^{(2)}) + \Sigma_3(D,s^{(3)}) + \Sigma_3^\vee(D,s^{(3)}) \\
\sim \;& \Sigma_2(D) + \Sigma_2^\vee(D) + \Sigma_3(D) + \Sigma_3^\vee(D) \\
= \;& \sum_k \frac{\frac{1}{2b_{2k}} + a_{2k}^2 + \frac{1}{2b_{3k}} + a_{3k}^2}{C_k + a_{1k}^2 + a_{2k}^2 + a_{3k}^2} \overset{(3.19)}{\sim} \sum_k \frac{\frac{1}{2b_{2k}} + a_{2k}^2 + \frac{1}{2b_{3k}} + a_{3k}^2}{\frac{1}{2b_{1k}} + a_{1k}^2} \\
=: \;& \Sigma_{23}^\vee(D) \sim \sum_k \frac{\frac{b_{1k}}{b_{2k}} + 2b_{1k}a_{2k}^2 + \frac{b_{1k}}{b_{3k}} + 2b_{1k}a_{3k}^2}{1 + 2b_{1k}a_{1k}^2}
\end{aligned}$$

$$\overset{(4.31)}{\sim} \sum_k \frac{a_{2k}^2 + a_{3k}^2}{1 + a_{1k}^2} =: \Sigma_{23}^a(D). \tag{4.61}$$

Remark 4.8. In Case (1.1.0) (resp., Case (1.2.0)), we have $\|Y_3\|^2 \sim \sum_{n\in\mathbb{Z}} a_{3n}^2 < \infty$ (resp.m $\|Y_2\|^2 \sim \sum_{n\in\mathbb{Z}} a_{2n}^2 < \infty$); therefore,

$$\Sigma_{23}^a(D) \sim \sum_k \frac{a_{2k}^2}{1 + a_{1k}^2} = \infty, \quad \text{resp.} \quad \Sigma_{23}^a(D) \sim \sum_k \frac{a_{3k}^2}{1 + a_{1k}^2} = \infty,$$

by Lemma 4.11, which contradicts (4.61). In Case (1.3.0), we have four possibilities:

(0) when $C_1 C_2 C_3 \ne 0$, $C_1 a_1 + C_2 a_2 + C_3 a_3 = h \in l_2(\mathbb{Z})$;
(1) when $C_1 = 0$, and hence $C_2 C_3 \ne 0$, $C_2 a_2 + C_3 a_3 = h \in l_2(\mathbb{Z})$;
(2) when $C_2 = 0$, and hence $C_1 C_3 \ne 0$, $C_1 a_1 + C_3 a_3 = h \in l_2(\mathbb{Z})$;
(3) when $C_3 = 0$, and hence $C_1 C_2 \ne 0$ $C_1 a_1 + C_2 a_2 = h \in l_2(\mathbb{Z})$.

In Case (0), we have $\Sigma_{23}^a(D) = \infty$ by Lemma 4.12, which contradicts (4.61). In Cases (2) and (3), we get $\Sigma_{23}^a(D) = \infty$ by Lemma 4.11, contradicting (4.61). Therefore, one of the expressions $\Sigma_2(D, s)$, $\Sigma_2^\vee(D, s)$, $\Sigma_3(D, s)$ or $\Sigma_3^\vee(D, s)$ is convergent. Hence, we can approximate D_{2n} or D_{3n}, and the proof is completed. Set $l_2 := l_2(\mathbb{Z})$. To study Case (1), we need the following lemma.

Lemma 4.13. *Let $C_2 Y_2 + C_3 Y_3 = h_{23} \in l_2$ for some $(C_2, C_3) \in \left(\mathbb{R} \setminus \{0\}\right)^2$ and $C_1 Y_1 + C_2 Y_2 \notin l_2$ or $C_1 Y_1 + C_3 Y_3 \notin l_2$ for all $(C_1, C_r) \in \left(\mathbb{R} \setminus \{0\}\right)^2$. Then,*

$$\Delta(Y_1, Y_2, Y_3) = \infty. \tag{4.62}$$

Proof. To prove (4.62), we have by (3.15),

$$
\begin{aligned}
\Delta(Y_1, Y_2, Y_3) \quad &= \quad \frac{\Gamma(Y_1) + \Gamma(Y_1, Y_2) + \Gamma(Y_1, Y_3) + \Gamma(Y_1, Y_2, Y_3)}{1 + \Gamma(Y_2) + \Gamma(Y_3) + \Gamma(Y_2, Y_3)} \\[2mm]
&\overset{(*)}{>} \quad \frac{\Gamma(Y_1, Y_2) + \Gamma(Y_1, Y_3)}{1 + (1 + c_2)\Gamma(Y_2) + \Gamma(Y_3)} \sim \frac{\Gamma(Y_1, Y_2) + \Gamma(Y_1, Y_3)}{\Gamma(Y_2) + \Gamma(Y_3)} \\[2mm]
&\overset{(4.64)}{\sim} \quad \frac{\Gamma(Y_1, Y_2) + \Gamma(Y_1, Y_3)}{2\Gamma(Y_2)} \\[2mm]
&\overset{(4.64)}{\sim} \quad \frac{\Gamma(Y_1, Y_2)}{\Gamma(Y_2)} + \frac{\Gamma(Y_1, Y_3)}{\Gamma(Y_3)} = \infty, \tag{4.63}
\end{aligned}
$$

$$\Gamma(Y_2) \sim \Gamma(Y_3), \quad \text{since } C_2 Y_2 + C_3 Y_3 = h \in l_2. \tag{4.64}$$

The relation $(*)$ holds by the inequality $\Gamma(Y_2, Y_3) \leq c_2 \Gamma(Y_2)$ since $C_2 Y_2 + C_3 Y_3 \in l_2$, and the relation (4.63) holds by Lemma 6.4 for $m = 2$. To prove (4.64), since $Y_2 \notin l_2$ and $h \in l_2$, we get

$$\frac{\Gamma(Y_3)}{\Gamma(Y_2)} = \frac{\|Y_3\|^2}{\|Y_2\|^2} = \frac{\|Y_2 + h\|^2}{\|Y_2\|^2} \leq \left(\frac{\|Y_2\| + \|h\|}{\|Y_2\|}\right)^2 = 1.$$

If $C_1 Y_1 + C_2 Y_2 \notin l_2$ for all $(C_1, C_2) \in \left(\mathbb{R} \setminus \{0\}\right)^2$, or $C_1 Y_1 + C_3 Y_3 \notin l_2$ for all $(C_1, C_3) \in \left(\mathbb{R} \setminus \{0\}\right)^2$, by Lemma 4.13, we get $\Delta(Y_1, Y_2, Y_3) = \infty$. Hence, we can approximate D_{1n} using Lemma 5.3, and the proof is completed. If $C_1 Y_1 + C_2 Y_2 = h_{12} \in l_2$ for some for $(C_1, C_2) \in \left(\mathbb{R} \setminus \{0\}\right)^2$ or $C_1 Y_1 + C_3 Y_3 = h_{13} \in l_2$ for some $(C_1, C_3) \in \left(\mathbb{R} \setminus \{0\}\right)^2$, then we have $h_{12} + \alpha h_{23} = C_1 Y_1 + C_2 Y_2 + C_3 Y_3 \in l_2$ or $h_{12} + \beta h_{13} = C_1 Y_1 + C_2 Y_2 + C_3 Y_3 \in l_2$, with $C_1 C_2 C_3 \neq 0$ for an appropriate $\alpha\beta \neq 0$, and we get Case (0). $\qquad\square$

4.4.6 *Case (c)*

In this case, both $\rho_2(C_2, -C_3)$ and $\rho_3(C_2, -C_3)$ are finite, i.e., we are in Case (2). Therefore, we cannot approximate $x_{2n}x_{2t}$, $x_{3n}x_{3t}$ by Lemma 5.1. By Lemma 4.9, $\Sigma_{23}(s) < \infty$, and hence $\Sigma_{23}(C_2, C_3) = \infty$. Indeed, reasoning as in Remark 3.8, we see that

$$\mu^{L_{\tau_{23}(\phi,s)}} \perp \mu, \quad \phi \in [0, 2\pi),$$

$$s > 0 \Leftrightarrow \Sigma_{23}(s) + \Sigma_{23}(C_2, C_3) = \infty, \quad s > 0, \qquad (4.65)$$

for $(C_2, C_3) \in \mathbb{R}^2 \setminus \{0\}$, where $\tau_{23}(\phi, s)$, $\Sigma_{23}(s)$ and $\Sigma_{23}(C_2, C_3)$ are defined as follows:

$$\tau_{12}(\phi, s) = \begin{pmatrix} \cos\phi & s^2 \sin\phi & 0 \\ s^{-2} \sin\phi & -\cos\phi & 0 \\ 0 & 0 & 1 \end{pmatrix}, \quad \tau_{23}(\phi, s) = \begin{pmatrix} 1 & 0 & 0 \\ 0 & \cos\phi & s^2 \sin\phi \\ 0 & s^{-2} \sin\phi & -\cos\phi \end{pmatrix},$$

$$(4.66)$$

$$\Sigma_{ij}(s) = \sum_{n \in \mathbb{Z}} \left(s^2 \sqrt{\frac{b_{in}}{b_{jn}}} - s^{-2} \sqrt{\frac{b_{jn}}{b_{in}}} \right)^2, \quad s \in \mathbb{R} \setminus \{0\}, \qquad (4.67)$$

$$\Sigma_{ij}(C_i, C_j) = \sum_{n \in \mathbb{Z}} (C_i^2 b_{in} + C_j^2 b_{jn})(C_i a_{in} + C_j a_{jn})^2. \qquad (4.68)$$

In this case, there are four possibilities for the pair $(\Sigma^{12}, \Sigma^{13})$:

(2.1) $(\Sigma^{12}, \Sigma^{13}) = (0, 0)$, i.e., $\Sigma^{12} < \infty$ and $\Sigma^{13} < \infty$;
(2.2) $(\Sigma^{12}, \Sigma^{13}) = (0, 1)$, i.e., $\Sigma^{12} < \infty$, but $\Sigma^{13} = \infty$;
(2.3) $(\Sigma^{12}, \Sigma^{13}) = (1, 0)$, i.e., $\Sigma^{12} = \infty$, but $\Sigma^{13} < \infty$;
(2.4) $(\Sigma^{12}, \Sigma^{13}) = (1, 1)$, i.e., $\Sigma^{12} = \infty$ and $\Sigma^{13} = \infty$.

Lemma 4.14. *In the case of (2.1), i.e., when $(\Sigma^{12}, \Sigma^{13}) = (0,0)$, we can approximate D_{rn} for $1 \leq r \leq 3$, and hence the representation is irreducible.*

Proof. Let $\Sigma^{12} < \infty$ and $\Sigma^{13} < \infty$. We have by (4.37),

$$\nu(C_1, C_2, C_3) \;\sim\; \sum_{k \in \mathbb{Z}} b_{1k}(C_1 a_{1k} + C_2 a_{2k} + C_3 a_{3k})^2$$

$$\overset{(2.1)}{\sim} \sum_{k \in \mathbb{Z}} \left[\frac{t^2}{4} \frac{b_{1k}}{b_{2k}} + \frac{s^2}{4} \frac{b_{1k}}{b_{3k}} + \frac{b_{1k}}{2}(-2a_{1k} + ta_{2k} + sa_{3k})^2 \right]$$

$$\overset{(3.47)}{=} S^L_{1,23}(\mu, t, s) = \infty.$$

Hence, D_{1n}, D_{2n}, D_{3n} η $\mathfrak{A}^3$, and the proof is finished. $\square$

Remark 4.9. Cases (2.2) and (2.3) do not occur.

Indeed, by Lemma 4.9, the three series $\Sigma_{23}(s)$, defined by (4.67), $\sum_{n \in \mathbb{Z}} c_n^2$ and $\sum_{n \in \mathbb{Z}} e_n^2$, are equivalent, where $\frac{s^4 b_{2n}}{b_{3n}} = (1 + c_n)$; see Lemma 4.10. In Case (c), we have $\sum_{n \in \mathbb{Z}} c_n^2 < \infty$; therefore, $\lim_n c_n = 0$, and hence $\lim_n d_n^{-1} = \lim_n \frac{b_{2n}}{b_{3n}} = s^{-4} > 0$. Recall that $d_n = \frac{d_{2n}}{d_{3n}} = \frac{b_{2n}}{b_{3n}}$. But this contradicts $(\Sigma^{12}, \Sigma^{13}) = (0, 1)$, or $(\Sigma^{12}, \Sigma^{13}) = (1, 0)$, since the two series

$$\Sigma^{12} = \sum_{n} d_{2n}^{-1} \quad \text{and} \quad \Sigma^{13} = \sum_{n} d_{3n}^{-1}$$

should be equivalent by $\lim_n \frac{d_{2n}}{d_{3n}} = s^{-4} > 0$. In **Case (2.4)**, we have

$$\Sigma_{23}(s) < \infty, \quad \Sigma_{23}(C_2, C_3) = \infty, \quad \Sigma^{12} = \Sigma^{13} = \infty. \tag{4.69}$$

To approximate D_{rn}, we need to estimate $\nu(C_1, C_2, C_3)$, defined by (4.36). By (4.37), we have

$$\nu(C_1, C_2, C_3) \sim \sum_{n \in \mathbb{Z}} b_{1n} \left(\sum_{r=1}^{3} C_r a_{rn} \right)^2.$$

Since $\|Y_1\|^2 = \sum_{n \in \mathbb{Z}} b_{1n} a_{1n}^2 \sim S_{11}^L(\mu) = \infty$, in Case (2.4), we have four possibilities for $y_{23} := (y_2, y_3) \in \{0, 1\}^2$:

	(2.4.1)	(2.4.2)	(2.4.3)	(2.4.4)
y_1	1	1	1	1
y_2	0	1	0	1
y_3	0	0	1	1

Remark 4.10. Cases (2.4.1)–(2.4.3) are not compatible with the condition $\Sigma_{23}(C_2, C_3) = \infty$ for all $(C_2, C_3) \in \mathbb{R}^2 \setminus \{0\}$.

So, it suffices to consider only **Case (2.4.4)** when $y_{123} = (1, 1, 1)$. Case (2.4.4) splits into two subcases:

(2.4.4.1) when $\Sigma_{12}(s_{12}) < \infty$ (resp., $\Sigma_{13}(s_{13}) < \infty$) for some $s_{12}, s_{13} > 0$;

(2.4.4.2) when both $\Sigma_{12}(s_{12}) = \Sigma_{13}(s_{13}) = \infty$ for all $s_{12}, s_{13} > 0$.

Case (2.4.4.1) does not occur. Indeed, we have here $\Sigma_{13}(s_{12}s_{23}) < \infty$ (resp., $\Sigma_{12}(s_{13}s_{23}^{-1}) < \infty$) since

$$\Sigma_{12}(s_{12}) < \infty \Leftrightarrow \mu_{(s_{12}^4 b_1, 0)} \sim \mu_{(b_2, 0)},$$

$$\Sigma_{23}(s_{23}) < \infty \Leftrightarrow \mu_{(s_{23}^4 b_2, 0)} \sim \mu_{(b_3, 0)},$$

where $\mu_{(b_r, 0)} = \otimes_{n \in \mathbb{Z}} \mu_{(b_{rk}, 0)}$ for $1 \leq r \leq 3$. Therefore,

$$\mu_{((s_{12}s_{23})^4 b_1, 0)} \sim \mu_{(b_3, 0)} \quad \Leftrightarrow \quad \Sigma_{13}(s_{12}s_{23}) < \infty.$$

Similarly, if $\Sigma_{13}(s_{13}) < \infty$ and $\Sigma_{23}(s_{23}) < \infty$, we have

$$\mu_{(s_{13}^4 b_1, 0)} \sim \mu_{(b_3, 0)}, \quad \mu_{(s_{23}^4 b_2, 0)} \sim \mu_{(b_3, 0)} \Rightarrow \mu_{((s_{13}s_{23}^{-1})^4 b_1, 0)} \sim \mu_{(b_2, 0)}.$$

Hence, $\Sigma_{12}(s_{13}s_{23}^{-1}) < \infty$. But condition $\Sigma_{13}(s_{12}s_{23}) + \Sigma_{12}(s_{13}s_{23}^{-1}) < \infty$ contradicts the first condition of (4.30). Indeed, we have by Lemma 4.10,

$$\Sigma_{12}(s) = \sum_{n \in \mathbb{Z}} \left(s^2 \sqrt{\frac{b_{1n}}{b_{2n}}} - s^{-2} \sqrt{\frac{b_{2n}}{b_{1n}}} \right)^2 \sim \sum_{n \in \mathbb{Z}} c_n^2 < \infty,$$

$$s^2 \sqrt{\frac{b_{1n}}{b_{2n}}} = 1 + c_n,$$

$$\Sigma_{13}(s) = \sum_{n \in \mathbb{Z}} \left(s^2 \sqrt{\frac{b_{1n}}{b_{3n}}} - s^{-2} \sqrt{\frac{b_{3n}}{b_{1n}}} \right)^2 \sim \sum_{n \in \mathbb{Z}} f_n^2 < \infty,$$

$$s^2 \sqrt{\frac{b_{1n}}{b_{3n}}} = 1 + f_n,$$

and $\lim_n c_n = \lim_n f_n = 0$. This contradicts $S_1(3) \sim \sum_n \frac{b_{1n}^2}{b_{2n} b_{3n}} < \infty$. Indeed,

$$\lim_{n \to \infty} \frac{b_{1n}^2}{b_{2n} b_{3n}} = s^{-4} \lim_{n \to \infty} (1 + c_n)^2 (1 + f_n)^2 = s^{-4} > 0.$$

Finally, to finish the case of $S = (0, 1, 1)$, we need to consider only Case (2.4.4.2) when $\Sigma_{12}(s_{12}) = \Sigma_{13}(s_{13}) = \infty$ for all $s_{12}, s_{13} > 0$.

By (4.30) and all the previous considerations, we have the following conditions:

$$S_1(3) \sim \sum_n \frac{b_{1n}^2}{b_{2n} b_{3n}} < \infty, \quad S_2(3) \sim \sum_n \frac{b_{2n}}{b_{3n}} = \infty,$$

$$S_3(3) \sim \sum_n \frac{b_{3n}}{b_{2n}} = \infty,$$

$$\Sigma_{23}(C_2, C_3) = \infty, \quad \Sigma^{12} = \sum_n \frac{b_{1n}}{b_{2n}} = \infty, \quad \Sigma^{13} = \sum_n \frac{b_{1n}}{b_{3n}} = \infty,$$

$$\tag{4.70}$$

$$\Sigma_{12}(s_{12}) = \Sigma_{13}(s_{13}) = \infty \quad \text{for all } s_{12}, s_{13} > 0,$$

$$\Sigma_{23}(s_{23}) < \infty \quad \text{for some } s_{23} > 0.$$

Remark 4.11. By (4.17) without loss of generality, we can suppose that (b_{1n}, b_{2n}, b_{3n}) is replaced with $(1, d_{2n}, d_{3n})$. Since $\Sigma_{23}(s) < \infty$, using notations (4.45) and (4.46) of Lemma 4.9,

$$\Sigma_{23}(s) = \sum_{n \in \mathbb{Z}} \left(\frac{s^2}{\sqrt{d_n}} - \frac{\sqrt{d_n}}{s^2} \right)^2 = \sum_{n \in \mathbb{Z}} \left(s^2 \sqrt{\frac{d_{2n}}{d_{3n}}} - s^{-2} \sqrt{\frac{d_{3n}}{d_{2n}}} \right)^2,$$

and taking into consideration (4.70), we can choose d_{2n} and d_{3n} as follows:

$$d_n = \frac{d_{3n}}{d_{2n}} = s^4(1 + c_n), \quad \sum_n c_n^2 < \infty, \quad \sum_n \frac{1}{d_{2n}^2} < \infty,$$

$$\sum_n \frac{1}{d_n} = \sum_n d_n = \infty. \tag{4.71}$$

Since $\sum_n c_n^2 < \infty$, we have $\sum_n \frac{b_{1n}^2}{b_{2n}b_{3n}} \sim \sum_n \frac{1}{d_{2n}^2}$, and the measures $\mu_{(d_3^{c,s},0)}$ and $\mu_{(d_3^s,0)}$ are equivalent, where

$$\mu_{(d_3^{c,s},0)} = \otimes_n \mu_{(s^4 d_{2n}(1+c_n),0)}, \quad \mu_{(d_3^s,0)} = \otimes_n \mu_{(s^4 d_{2n},0)}.$$

Hence, we can choose $c_n \equiv 0$ and $s = 1$. So, to complete the case of $S = (0,1,1)$, we should prove irreducibility for $b = (1, d_{2n}, d_{2n})_{n \in \mathbb{Z}}$ with the only condition:

$$\sum_n d_{2n}^{-2} < \infty. \quad \text{Since } d_n \equiv 1, \quad \text{we have } \sum_n d_n^{-1} = \sum_n d_n = \infty. \tag{4.72}$$

Example 4.2. The pairwise conditions

$$\|C_r Y_r + C_s Y_s\|^2 = \infty \quad \text{for } 1 \le r < s \le 3 \quad \text{do not imply}$$

$$\left\| \sum_{r=1}^3 C_r Y_r \right\|^2 = \infty.$$

Let $a_{r,n} = a_{r,-n}$ for $n \in \mathbb{N}$ and $a_{1,0} = 1$, $a_{2,0} = 2$, $a_{3,0} = 3$. We define $a_{r,n}$ for $n \in \mathbb{N}$ as follows:

$$a_{1n} = \begin{cases} 2, & n = 2k+1, \\ 1, & n = 2k, \end{cases} \quad a_{2n} = \begin{cases} 1, & n = 2k+1, \\ 2, & n = 2k, \end{cases} \quad a_{3n} \equiv 3. \tag{4.73}$$

Then, we clearly have, for arbitrary $(C_1, C_2, C_3) \in \mathbb{R}^3 \setminus \{0\}$,

$$\|C_1 a_1 + C_2 a_2\|^2 = \infty, \quad \|C_1 a_1 + C_3 a_3\|^2 = \infty, \quad \|C_2 a_2 + C_3 a_3\|^2 = \infty, \tag{4.74}$$

$$\text{but } a_1 + a_2 - a_3 = 0; \quad \text{hence, } \|a_1 + a_2 - a_3\|^2 = 0. \tag{4.75}$$

Example 4.3. Let us consider the measure $\mu^3_{(b,a)}$ with $a = \left(a_{rn}\right)_{r,n}$ from Example 4.2 and $b = (b_{1n}, b_{2n}, b_{3n})$, defined as follows:

$$b_{1n} \equiv 1, \quad d_{2n} = d_{3n} = |n| \quad \text{for } n \in \mathbb{Z} \setminus \{0\}, \quad d_{20} = d_{30} = 1. \quad (4.76)$$

Lemma 4.15. *In Example* 4.2, *we have (for $n \in \mathbb{N}$ only)*,

$$\Delta(a_1, a_2, a_3) = 2, \quad \Delta(a_2, a_3, a_1) = 2, \quad \Delta(a_3, a_1, a_2) = 2, \quad (4.77)$$

where $a_r = (a_{rn})_{n \in \mathbb{N}}, \ 1 \leq r \leq 3$.

Proof. Set $a_r(n) = (a_{rl})_{l=1}^n$ for $1 \leq r \leq 3$ and $n \in \mathbb{N}$. Then, for $1 \leq k < r \leq 3$,

$$\Gamma(a_k(n)) \sim \Gamma(a_1(n) + a_2(n)) \sim n, \quad \Gamma(a_k(n), a_r(n)) \sim \frac{n(n-1)}{2},$$

$$\Gamma(a_1, a_2, a_3) = 0.$$

We observe that $\Gamma(a_k, a_k + a_r) = \Gamma(a_k, a_r)$ for $k \neq r$. Since $a_3 = a_1 + a_2$, we get

$$\Delta(a_1, a_2, a_3) = \frac{\Gamma(a_1) + \Gamma(a_1, a_2) + \Gamma(a_1, a_3) + \Gamma(a_1, a_2, a_3)}{1 + \Gamma(a_2) + \Gamma(a_3) + \Gamma(a_2, a_3)}$$

$$= \frac{\Gamma(a_1) + \Gamma(a_1, a_2) + \Gamma(a_1, a_1 + a_2) + \Gamma(a_1, a_2, a_1 + a_2)}{1 + \Gamma(a_2) + \Gamma(a_1 + a_2) + \Gamma(a_2, a_1 + a_2)}$$

$$= \frac{\Gamma(a_1) + 2\Gamma(a_1, a_2)}{1 + \Gamma(a_2) + \Gamma(a_1 + a_2) + \Gamma(a_1, a_2)} = 2,$$

$$\Delta(a_2, a_3, a_1) = \frac{\Gamma(a_2) + \Gamma(a_2, a_3) + \Gamma(a_2, a_1) + \Gamma(a_2, a_3, a_1)}{1 + \Gamma(a_3) + \Gamma(a_1) + \Gamma(a_3, a_1)}$$

$$= \frac{\Gamma(a_2) + \Gamma(a_2, a_1 + a_2) + \Gamma(a_2, a_1) + \Gamma(a_2, a_1 + a_2, a_1)}{1 + \Gamma(a_1 + a_2) + \Gamma(a_1) + \Gamma(a_1 + a_2, a_1)}$$

$$= \frac{\Gamma(a_2) + 2\Gamma(a_2, a_1)}{1 + \Gamma(a_1 + a_2) + \Gamma(a_1) + \Gamma(a_2, a_1)} = 2,$$

$$\Delta(a_3, a_1, a_2) = \frac{\Gamma(a_3) + \Gamma(a_3, a_1) + \Gamma(a_3, a_2) + \Gamma(a_3, a_1, a_2)}{1 + \Gamma(a_1) + \Gamma(a_2) + \Gamma(a_1, a_2)}$$

$$\begin{aligned}
&= \frac{\Gamma(a_1+a_2)+\Gamma(a_1+a_2,a_1)+\Gamma(a_1+a_2,a_2)}{1+\Gamma(a_1)+\Gamma(a_2)+\Gamma(a_1,a_2)} \\[2pt]
&\quad\quad \frac{+\Gamma(a_1+a_2,a_1,a_2)}{} \\[4pt]
&= \frac{\Gamma(a_1+a_2)+2\Gamma(a_1,a_2)}{1+\Gamma(a_1)+\Gamma(a_2)+\Gamma(a_1,a_2)} = 2.
\end{aligned}$$

We use two facts for $1 \leq r \leq 2$:

$$\frac{\Gamma(a_1,a_2)}{\Gamma(a_r)} = \infty \quad \text{and} \quad \Gamma(a_1+a_2) \leq \Gamma(a_1)+\Gamma(a_2)+2\sqrt{\Gamma(a_1)\Gamma(a_2)}.$$

The first relation follows from Lemma 6.4 for $m = 2$ since $\|C_1 a_1 + C_2 a_2\|^2 = \infty$. We get

$$\frac{\Gamma(a_1,a_2)}{\Gamma(a_r)} = \lim_{n\to\infty} \frac{\Gamma(a_1(n),a_2(n))}{\Gamma(a_r(n))} = \infty.$$

Recall that $\Gamma(a) = \|a\|^2$. The inequality follows from $\|a_1+a_2\| \leq \|a_1\| + \|a_2\|$, i.e., $\sqrt{\Gamma(a_1+a_2)} \leq \sqrt{\Gamma(a_1)} + \sqrt{\Gamma(a_2)}$. $\square$

By Lemma 3.5, we have

$$(\mu^3_{(b,a)})^{L_t} \perp \mu^3_{(b,a)} \Leftrightarrow \Sigma^{\pm}(t) := \Sigma^{\pm}_1(t) + \Sigma_2(t) = \infty,$$

where $\Sigma^{\pm}_1(t)$ and $\Sigma_2(t^{-1})$ are defined by (3.37) and (3.34). In Example 4.3, we cannot approximate x_{2n}, x_{3n} since in this case, we have

$$\Delta(Y_2^{(2)}, Y_3^{(2)}) = 1, \quad \Delta(Y_3^{(3)}, Y_2^{(3)}) = 1. \tag{4.78}$$

Indeed, by (4.34), we have

$$\Delta(Y_2^{(2)}, Y_3^{(2)}) = \frac{\Gamma(Y_2^{(2)})+\Gamma(Y_2^{(2)},Y_3^{(2)})}{1+\Gamma(Y_3^{(2)})},$$

$$\Delta(Y_3^{(3)}, Y_2^{(3)}) = \frac{\Gamma(Y_3^{(3)})+\Gamma(Y_3^{(3)},Y_2^{(3)})}{1+\Gamma(Y_2^{(3)})}.$$

In Example 4.3, we have $d_n = \frac{d_{3n}}{d_{2n}} \equiv 1$, and hence, by (4.41), we have

$$\|Y_2^{(2)}\|^2 \sim \sum_{n\in\mathbb{Z}} \frac{1}{1+2d_n} = \sum_{n\in\mathbb{Z}} \frac{1}{3}, \quad \|Y_3^{(2)}\|^2 \sim \sum_{n\in\mathbb{Z}} \frac{d_n^2}{1+2d_n} = \sum_{n\in\mathbb{Z}} \frac{1}{3},$$

$$\|Y_2^{(3)}\|^2 \sim \sum_{n\in\mathbb{Z}} \frac{1}{d_n^2 + 2d_n} = \sum_{n\in\mathbb{Z}} \frac{1}{3}, \quad \|Y_3^{(3)}\|^2 \sim \sum_{n\in\mathbb{Z}} \frac{d_n^2}{d_n^2 + 2d_n} = \sum_{n\in\mathbb{Z}} \frac{1}{3}.$$

Therefore, $\Gamma(Y_2^{(2)}, Y_3^{(2)}) = \Gamma(Y_3^{(3)}, Y_2^{(3)}) = 0$, and

$$\Delta(Y_2^{(2)}, Y_3^{(2)}) = \frac{\Gamma(Y_2^{(2)})}{1 + \Gamma(Y_3^{(2)})} = 1, \quad \Delta(Y_3^{(3)}, Y_2^{(3)}) = \frac{\Gamma(Y_3^{(3)})}{1 + \Gamma(Y_2^{(3)})} = 1.$$

Since $b_{1n} \equiv 1$, by (4.42), we get

$$\|Y_1\|^2 \sim \sum_{n\in\mathbb{Z}} a_{1n}^2, \quad \|Y_2\|^2 \sim \sum_{n\in\mathbb{Z}} a_{2n}^2, \quad \|Y_3\|^2 \sim \sum_{n\in\mathbb{Z}} a_{3n}^2,$$

so we have

$$\nu(C_1, C_2, C_3) \overset{(4.37)}{\sim} \sum_{n\in\mathbb{Z}} b_{1n} \left(\sum_{r=1}^{3} C_r a_{rn} \right)^2 = \sum_{n\in\mathbb{Z}} \left(\sum_{r=1}^{3} C_r a_{rn} \right)^2.$$

However, in Example 4.2, there does not exist $t \in \pm\mathrm{SL}(3,\mathbb{R}) \setminus \{e\}$ such that $\nu(C_1, C_2, C_3) = \infty$ for all $(C_1, C_2, C_3) \in \mathbb{R}^3 \setminus \{0\}$ to approximate some D_{rn}.

4.4.7 *Approximations of $x_{2k}x_{2r} + x_{3k}x_{3r}$ in Case (c)*

Since we cannot approximate $x_{2n}x_{2r}$, $x_{3n}x_{3r}$ using Lemma 5.1 in Case (c), we shall try to approximate $x_{2k}x_{2r} + s^4 x_{3k}x_{3r}$ by an appropriate combination of $A_{kn}A_{rn}$ for $n \in \mathbb{Z}$. Let $s = 1$, and the general case is similar.

Lemma 4.16. *For any $k, r \in \mathbb{Z}$, one has*

$$(x_{2k}x_{2r} + x_{3k}x_{3r})\mathbf{1} \in \langle A_{kn}A_{rn}\mathbf{1} \mid n \in \mathbb{Z}\rangle \Leftrightarrow \Delta(Y^{(2)}, Y^{(1)}) = \infty,$$

$$(4.79)$$

where $Y^{(r)} = \left(\dfrac{b_{rn}}{\sqrt{\lambda_n}} \right)_{n\in\mathbb{Z}}$, $1 \le r \le 2$, and $\lambda_n = (b_{1n} + b_{2n} + b_{3n})^2 - b_{1n}^2$.

Proof. The proof of Lemma 4.16 is based on that of Lemma 6.6 for $m = 1$. We study when $(x_{2k}x_{2r} + x_{3k}x_{3r})\mathbf{1} \in \langle A_{kn}A_{rn}\mathbf{1} \mid n \in \mathbb{Z}\rangle$. Since

$$A_{kn}A_{rn} = (x_{1k}D_{1n} + x_{2k}D_{2n} + x_{3k}D_{3n})(x_{1r}D_{1n} + x_{2r}D_{2n} + x_{3r}D_{3n})$$

$$= x_{1k}x_{1r}D_{1n}^2 + x_{2k}x_{2r}D_{2n}^2 + x_{3k}x_{3r}D_{3n}^2 + (x_{1k}x_{2r} + x_{2k}x_{1r})$$

$$\times D_{1n}D_{2n} + (x_{1k}x_{3r} + x_{3k}x_{1r})D_{1n}D_{3n}$$

$$+ (x_{2k}x_{3r} + x_{3k}x_{2r})D_{2n}D_{3n},$$

and $MD_{rn}^2\mathbf{1} = -\frac{b_{rn}}{2}$, for $2 \leq r \leq 3$, we take $t = (t_n)_{n=-m}^m$ as follows:

$$(t, b_2) = (t, b_3) = 1,$$

where $t = (t_n)_{k=-m}^m, \quad b_2 = -\left(\frac{b_{2n}}{2}\right)_{n=-m}^m, \quad b_3 = -\left(\frac{b_{3n}}{2}\right)_{n=-m}^m.$

We have

$$\left\|\left[\sum_{n=-m}^m t_n A_{kn}A_{rn} - (x_{2k}x_{2r} + x_{3k}x_{3r})\right]\mathbf{1}\right\|^2$$

$$= \left\|\sum_{n=-m}^m t_n\left[x_{1k}x_{1r}D_{1n}^2 + x_{2k}x_{2r}\left(D_{2n}^2 + \frac{b_{2n}}{2}\right)\right.\right.$$

$$+ x_{3k}x_{3r}\left(D_{3n}^2 + \frac{b_{3n}}{2}\right) + (x_{1k}x_{2r} + x_{2k}x_{1r})D_{1n}D_{2n}$$

$$\left.\left. + (x_{1k}x_{3r} + x_{3k}x_{1r})D_{1n}D_{3n} + (x_{2k}x_{3r} + x_{3k}x_{2r})D_{2n}D_{3n}\right]\mathbf{1}\right\|^2$$

$$= \sum_{-m \leq n, l \leq m} (f_n, f_l)t_n t_l =: (A_{2m+1}t, t), \tag{4.80}$$

where $A_{2m+1} = (f_n, f_l)_{n,l=-m}^m$,

$$f_n = \sum_{i=1}^3 f_n^i + \sum_{1 \leq i < j \leq 3} f_n^{ij}, \tag{4.81}$$

with $f_n^i = x_{ik}x_{ir}\left(D_{in}^2 + \frac{b_{in}}{2}(1 - \delta_{i1})\right)\mathbf{1}, \quad f_n^{ij} = (x_{ik}x_{jr} + x_{jk}x_{ir})D_{in}D_{jn}\mathbf{1},$

for $1 \leq i \leq 3$, $1 \leq i < j \leq 3$. Since $f_n^{i'} \perp f_n^{ij}$, $\quad f_n^{ij} \perp f_n^{i'j'}$ for different (ij), $(i'j')$, writing $c_{kn} = \|x_{kn}\|^2 = \frac{1}{2b_{kn}} + a_{kn}^2$, we get

$$(f_n, f_n) = \sum_{i=1}^{3} \|f_n^i\|^2 + \sum_{1 \leq i < j \leq 3} \|f_n^{ij}\|^2$$

$$= c_{1k}c_{1r}3\left(\frac{b_{1n}}{2}\right)^2 + c_{2k}c_{2r}2\left(\frac{b_{2n}}{2}\right)^2 + c_{3k}c_{3r}2\left(\frac{b_{3n}}{2}\right)^2$$

$$+ (c_{1k}c_{2r} + c_{2k}c_{1r} + 2a_{1k}a_{2r}a_{2k}a_{1r})\frac{b_{1n}}{2}\frac{b_{2n}}{2}$$

$$+ (c_{1k}c_{3r} + c_{3k}c_{1r} + 2a_{1k}a_{3r}a_{3k}a_{1r})\frac{b_{1n}}{2}\frac{b_{3n}}{2}$$

$$+ (c_{2k}c_{3r} + c_{3k}c_{2r} + 2a_{2k}a_{3r}a_{3k}a_{2r})\frac{b_{2n}}{2}\frac{b_{3n}}{2}$$

$$\sim (b_{1n} + b_{2n} + b_{3n})^2,$$

$$(f_n, f_l) = (f_n^1, f_l^1) = c_{1k}c_{1r}\frac{b_{1n}}{2}\frac{b_{1l}}{2} \sim b_{1n}b_{1l}.$$

Finally, we get

$$(f_n, f_n) \sim (b_{1n} + b_{2n} + b_{3n})^2, \quad (f_n, f_l) \sim b_{1n}b_{1l}, \quad n \neq l. \tag{4.82}$$

Set

$$\lambda_n = (b_{1n} + b_{2n} + b_{3n})^2 - b_{1n}^2, \quad g_n = (b_{1n}), \tag{4.83}$$

then

$$(f_n, f_n) \sim \lambda_n + (g_n, g_n), \quad (f_n, f_l) \sim (g_n, g_l). \tag{4.84}$$

For $A_{2m+1} = \left((f_n, f_l)\right)_{n,l=-m}^{m}$ and $b_2 = b_3 = -(b_{2n}/2)_{n=-m}^{m} \in \mathbb{R}^{2m+1}$, we have

$$A_{2m+1} = \sum_{n=-m}^{m} \lambda_n E_{nn} + \gamma(g_{-m}, \ldots, g_0, \ldots, g_m).$$

To complete the proof, it suffices to use Lemma 6.6 for $m = 1$. $\qquad \square$

Remark 4.12. In Case (c), we can approximate $x_{2k}x_{2r} + x_{3k}x_{3r}$ since $\Delta(Y^{(2)}, Y^{(1)}) = \infty$.

By (4.79), we have

$$\Delta(Y^{(2)}, Y^{(1)}) = \frac{\Gamma(Y^{(2)}) + \Gamma(Y^{(2)}, Y^{(1)})}{1 + \Gamma(Y^{(1)})} > \frac{\Gamma(Y^{(2)})}{1 + \Gamma(Y^{(1)})} = \infty$$

since $\Gamma(Y^{(2)}) = \infty$ by $\sum_n \frac{1}{d_{2n}^2} < \infty$ and $\Gamma(Y^{(1)}) < \infty$. Indeed,

$$\Gamma(Y^{(2)}) \;=\; \sum_{n \in \mathbb{Z}} \frac{b_{2n}^2}{\lambda_n} = \sum_{n \in \mathbb{Z}} \frac{d_{2n}^2}{(1 + 2d_{2n})^2 - 1} \sim \sum_{n \in \mathbb{Z}} \frac{d_{2n}^2}{d_{2n} + d_{2n}^2}$$

$$\overset{(3.19)}{\sim} \sum_{n \in \mathbb{Z}} d_{2n} = \infty,$$

$$\Gamma(Y^{(1)}) \overset{(4.79)}{=} \sum_{n \in \mathbb{Z}} \frac{1}{(1 + 2d_{2n})^2 - 1} \sim \sum_{n \in \mathbb{Z}} \frac{1}{d_{2n} + d_{2n}^2}$$

$$< \sum_n \frac{1}{d_{2n}^2} \overset{(4.71)}{<} \infty.$$

Lemma 4.17. *Let $\{r, s\}$ be a cyclic permutation of $\{2, 3\}$. Then, for all $k \in \mathbb{Z}$,*

$$x_{rk}\mathbf{1} \in \langle (x_{2k}x_{2n} + x_{3k}x_{3n})\mathbf{1} \mid n \in \mathbb{Z} \rangle \Leftrightarrow \sigma_r(\mu) = \sum_{n \in \mathbb{Z}} \frac{a_{rn}^2}{\frac{1}{2b_{rn}} + c_{sn}^2} = \infty.$$

$$(4.85)$$

Proof. Recall the notation $c_{rn} = \frac{1}{2b_{rn}} + a_{rn}^2$. Since $M x_{2n}\mathbf{1} = a_{2n}$, we take $t = (t_n)_{n=-m}^{m}$ as follows: $(t, a_2) = 1$, where $a_2 = (a_{2n})_{n=-m}^{m}$. For $r = 2$, we get

$$\left\| \left[\sum_{n=-m}^{m} t_n \big(x_{2k}x_{2n} + x_{3k}x_{3n} \big) - x_{2k}x_{2n} \right] \mathbf{1} \right\|^2$$

$$= \left\| \left[\sum_{n=-m}^{m} t_n \big(x_{2k}(x_{2n} - a_{2n}) + x_{3k}x_{3n} \big) \right] \mathbf{1} \right\|^2$$

$$= \|x_{2k}\mathbf{1}\|^2 \left\| \sum_{n=-m}^{m} t_n(x_{2n} - a_{2n})\mathbf{1} \right\|^2 + \|x_{3k}\mathbf{1}\|^2 \left\| \sum_{n=-m}^{m} t_n x_{3n} \mathbf{1} \right\|^2$$

$$= c_{2k} \sum_{n=-m}^{m} t_n^2 \frac{1}{2b_{2n}} + c_{3k} \sum_{n=-m}^{m} t_n^2 c_{3n} \sim \sum_{n=-m}^{m} t_n^2 \left(\frac{1}{2b_{2n}} + c_{3n}^2 \right).$$

By (6.3), we get (4.85). $\qquad\square$

Define

$$\sigma_2(\mu) = \sum_{n\in\mathbb{Z}} \frac{a_{2n}^2}{\frac{1}{2b_{2n}} + \frac{1}{2b_{3n}} + a_{3n}^2}, \quad \sigma_3(\mu) = \sum_{\in\mathbb{Z}} \frac{a_{3n}^2}{\frac{1}{2b_{2n}} + \frac{1}{2b_{3n}} + a_{2n}^2}.$$

Remark 4.13. Suppose $\sigma_2(\mu) + \sigma_3(\mu) < \infty$, which contradicts $\Sigma_{23}(C_2, C_3) = \infty$ for $(C_2, C_3) \in \mathbb{R}^2 \setminus \{0\}$, where $\Sigma_{23}(C_2, C_3)$ is defined by (4.68).

Proof. Indeed, we have

$$\infty > \sigma_2(\mu) + \sigma_3(\mu) = \sum_{n\in\mathbb{Z}} \frac{a_{2n}^2}{\frac{1}{2b_{2n}} + \frac{1}{2b_{3n}} + a_{3n}^2} + \sum_{n\in\mathbb{Z}} \frac{a_{3n}^2}{\frac{1}{2b_{2n}} + \frac{1}{2b_{3n}} + a_{2n}^2}$$

$$\sim \sum_{n\in\mathbb{Z}} \frac{a_{2n}^2 + a_{3n}^2}{\frac{1}{2b_{2n}} + \frac{1}{2b_{3n}} + a_{2n}^2 + a_{3n}^2} \sim \sum_{n\in\mathbb{Z}} \frac{a_{2n}^2 + a_{3n}^2}{\frac{1}{2b_{2n}} + \frac{1}{2b_{3n}}}$$

$$\overset{(4.71)}{=} \frac{2}{1 + s^{-4}} \sum_{n\in\mathbb{Z}} b_{2n}(a_{2n}^2 + a_{3n}^2).$$

This contradicts $\Sigma_{23}(C_2, C_3) = \infty$. Indeed, by $b_{3n} = s^4 b_{2n}$, referring to (4.71), we have

$$\Sigma_{23}(C_2, C_3) = \sum_{n\in\mathbb{Z}} (C_2^2 + C_3^2 s^4) b_{2n} (C_2 a_{2n} + C_3 a_{3n})^2 < \infty. \qquad\square$$

Finally, we have $\sigma_2(\mu) + \sigma_3(\mu) = \infty$; therefore, we have $x_{rn} \eta \, \mathfrak{A}^3$ for some $2 \leq r \leq 3$. Let $x_{3n} \eta \, \mathfrak{A}^3$. Then, we can approximate x_{2n} by combinations of $x_{2n}x_{2k}$, $k \in \mathbb{Z}$ using an analogue of Lemma 4.4. To approximate

D_{rn}, $1 \leq r \leq 3$, we again proceed as in Section 4.4.4. As in (4.51), we get

$$\|Y_1\|^2 \sim \sum_{n \in \mathbb{Z}} a_{1n}^2, \quad \|Y_2\|^2 \sim \sum_{n \in \mathbb{Z}} a_{2n}^2, \quad \|Y_3\|^2 \sim \sum_{n \in \mathbb{Z}} a_{3n}^2.$$

Indeed, for example, by (4.5), we get

$$\|Y_1\|^2 = \sum_{n \in \mathbb{Z}} \frac{a_{1n}^2}{\frac{1}{2b_{1n}} + \frac{1}{2b_{2n}} + \frac{1}{2b_{3n}}} = \sum_{n \in \mathbb{Z}} \frac{a_{1n}^2}{\frac{1}{2} + \frac{1}{d_{2n}}} \overset{(4.72)}{\sim} \sum_{n \in \mathbb{Z}} a_{1n}^2.$$

Again, as in (4.52), we have four possibilities: (1.0), (1.1), (1.2) and (1.3). The corresponding expressions in (4.56), (4.57) and (4.58) become as follows:

$$\nu_{12}(C_1, C_2) := \|C_1 Y_1 + C_2 Y_2\|^2 \sim \sum_{n \in \mathbb{Z}} (C_1 a_{1n} + C_2 a_{2n})^2,$$

$$\nu_{13}(C_1, C_3) := \|C_1 Y_1 + C_3 Y_3\|^2 \sim \sum_{n \in \mathbb{Z}} (C_1 a_{1n} + C_3 a_{3n})^2,$$

$$\nu(C_1, C_2, C_3) = \sum_{n \in \mathbb{Z}} (C_1 a_{1n} + C_2 a_{2n} + C_3 a_{3n})^2.$$

To study Cases (1.1.1)–(1.3.1), we should use Remark 4.5. We can approximate in these cases D_{1n} and D_{2n} in (4.53), D_{1n} and D_{3n} in (4.54) and D_{1n}, D_{2n}, D_{3n} in (4.55). The proof of irreducibility is completed in these cases because we have D_{rn}, x_{2n}, $x_{3n} \, \eta \, \mathfrak{A}^3$ for some $1 \leq r \leq 3$. Following Remark 4.7, we can use Lemma 5.4 and its analogue to approximate D_{2n} and D_{3n} with the corresponding expressions $\Sigma_2(D, s)$, $\Sigma_2^\vee(D, s)$ and $\Sigma_3(D), \Sigma_3^\vee(D)$. If one of the expressions $\Sigma_2(D, s)$, $\Sigma_2^\vee(D, s)$, $\Sigma_3(D, s)$ or $\Sigma_3^\vee(D, s)$ is divergent for some sequence $s = (s_k)_{k \in \mathbb{Z}}$, we can approximate D_{2k} or D_{3k}, and the proof is completed. Suppose that for any sequence $s = (s_k)_{k \in \mathbb{Z}}$, we have

$$\Sigma_2(D, s) + \Sigma_2^\vee(D, s) + \Sigma_3(D, s) + \Sigma_3^\vee(D, s) < \infty.$$

Then, by (4.61), we have

$$\infty > \Sigma_{23}^{\vee}(D) \;=\; \sum_k \frac{\frac{1}{2b_{2k}} + a_{2k}^2 + \frac{1}{2b_{3k}} + a_{3k}^2}{\frac{1}{2b_{1k}} + a_{1k}^2} = \sum_k \frac{\frac{1}{d_{2k}} + a_{2k}^2 + a_{3k}^2}{\frac{1}{2} + a_{1k}^2}$$

$$\overset{(4.72)}{\sim} \sum_k \frac{a_{2k}^2 + a_{3k}^2}{1 + a_{1k}^2} =: \Sigma_{23}^a(D).$$

To study Cases (1.1.0)–(1.3.0), we should follow Remark 4.8.

4.5   Case of $S = (1, 1, 1)$

Denote by

$$\Sigma_{123}(s) = (\Sigma_{12}(s_1), \Sigma_{23}(s_2), \Sigma_{13}(s_3)), \tag{4.86}$$

where $s = (s_1, s_2, s_3)$ and $\Sigma_{ij}(s)$ are defined by (4.67) for $1 \le i < j \le 3$. In terms of Remark 4.2, we have 2^3 possibilities for $\Sigma_{123}(s) \in \{0, 1\}^3$:

	(0)	(1)	(2)	(3)	(4)	(5)	(6)	(7)
$\Sigma_{12}(s_1)$	0	0	0	0	1	1	1	1
$\Sigma_{23}(s_2)$	0	0	1	1	0	0	1	1
$\Sigma_{13}(s_3)$	0	1	0	1	0	1	0	1

Cases (1), (2) and (4) and, respectively, Cases (3), (5) and (6) are obtained as the cyclic permutations of three measures, $\mu^{(1)}, \mu^{(2)}, \mu^{(3)}$, defined as follows:

$$\mu^{(r)} = \otimes_{n \in \mathbb{Z}} \mu_{(b_{rn}, a_{rn})}, \quad 1 \le r \le 3, \quad \mu_0^{(r)} = \otimes_{n \in \mathbb{Z}} \mu_{(b_{rn}, 0)}, \quad 1 \le r \le 3.$$

$$\tag{4.87}$$

Cases (1), (2) and (4) cannot be realised. We prove this only for Case (1). By Lemma 6.7, we have $\Sigma_{12}(s_1) < \infty \;\Leftrightarrow\; \mu_0^{(1)} \sim \mu_0^{(2)}$ and $\Sigma_{23}(s_2) < \infty \Leftrightarrow \mu_0^{(2)} \sim \mu_0^{(3)}$. Hence, $\mu_0^{(1)} \sim \mu_0^{(3)}$, which contradicts $\Sigma_{13}(s_2) = \infty \Leftrightarrow \mu_0^{(1)} \perp \mu_0^{(3)}$. Finally, we are left with the three cases (0), (3) and (7):

Case (0), i.e., $\Sigma_{123}(s) = (0, 0, 0)$;

Case (3), i.e., $\Sigma_{123}(s) = (0, 1, 1)$;

Case (7), i.e., $\Sigma_{123}(s) = (1, 1, 1)$.

4.5.1   *Case $\Sigma_{123}(s) = (0,0,0)$*

In Case (0), we have for some $s = (s_1, s_2, s_3) \in (\mathbb{R}_+)^3$,

$$\Sigma_{12}(s_1) < \infty, \quad \Sigma_{23}(s_2) < \infty, \quad \Sigma_{13}(s_3) < \infty.$$

In this case, we get $\mu_0^{(1)} \sim \mu_0^{(2)} \sim \mu_0^{(3)}$. By (4.17), we can make the following change of variables:

$$\begin{pmatrix} b_{1n}\; b_{2n}\; b_{3n} \\ a_{1n}\; a_{2n}\; a_{3n} \end{pmatrix} \to \begin{pmatrix} b'_{1n}\; b'_{2n}\; b'_{3n} \\ a'_{1n}\; a'_{2n}\; a'_{3n} \end{pmatrix} = \begin{pmatrix} 1 & \frac{b_{2n}}{b_{1n}} & \frac{b_{3n}}{b_{1n}} \\ a_{1n}\sqrt{b_{1n}} & a_{2n}\sqrt{b_{1n}} & a_{3n}\sqrt{b_{1n}} \end{pmatrix}.$$

Remark 4.14. By Lemma 6.8, we can suppose that

$$b = (b_{1n}, b_{2n}, b_{3n})_{n \in \mathbb{Z}} = (1, 1 + c_n, 1 + e_n)_{n \in \mathbb{Z}},$$

$$\sum_n c_n^2 < \infty, \quad \sum_n e_n^2 < \infty. \tag{4.88}$$

But the two measures $\mu_{(b,a)}$ and $\mu_{(\mathbb{I},a)}$ are equivalent, where b is defined by (4.88) and

$$\mathbb{I} := (1, 1, 1)_{n \in \mathbb{Z}}. \tag{4.89}$$

Finally, it is sufficient to consider the measure $\mu_{(\mathbb{I},a)}$.

Example 4.4. Let $b_{1n} = b_{2n} = b_{3n} \equiv 1$, $n \in \mathbb{Z}$.

(a) Take $a_n = (a_{1n}, a_{2n}, a_{3n})$, $n \in \mathbb{Z}$, as was defined in Example 4.2:

$$a_{1n} = \begin{cases} 2, & \text{if } n = 2k+1, \\ 1, & \text{if } n = 2k; \end{cases} \qquad a_{2n} = \begin{cases} 1, & \text{if } n = 2k+1, \\ 2, & \text{if } n = 2k; \end{cases} \qquad a_{3n} \equiv 3.$$

Then, $a_1 + a_2 - a_3 = 0$, where $a_r = (a_{rn})_{n \in \mathbb{Z}}$.

(b) Take any $a_r = (a_{rn})_{n \in \mathbb{Z}}$ such that $a_1, a_2, a_3 \notin l_2(\mathbb{Z})$, but $C_1 a_1 + C_2 a_2 + C_3 a_3 \in l_2(\mathbb{Z})$ for some $(C_1, C_2, C_3) \in \mathbb{R}^3 \setminus \{0\}$.

Example 4.5. Let $b_{1n} = b_{2n} = b_{3n} \equiv 1$, $n \in \mathbb{Z}$ and $a = (a_{1n}, a_{2n}, a_{3n})_{n \in \mathbb{Z}}$ such that $a_1, a_2, a_3 \notin l_2(\mathbb{Z})$, but the measure $\mu_{(b,a)}^3$ satisfies the orthogonality conditions. The case $\Sigma_{123}(s) = (0,0,0)$ is reduced to this example.

Remark 4.15. Since the measure $\mu^3_{(b,0)}$ is *standard* in Examples 4.4 and 4.5, i.e., it is invariant under rotations $\pm O(3)$, we have

$$(\mu^3_{(b,0)})^{L_t} = \mu^3_{(b,0)} \quad \text{for all } t \in \pm O(3). \tag{4.90}$$

By Lemma 3.6, the orthogonality condition $(\mu^3_{(b,a)})^{L_t} \perp \mu^3_{(b,a)}$ for $t \in \pm O(3) \setminus \{e\}$ is equivalent to

$$\Sigma_1^{\pm}(t) + \Sigma_2(t) = \infty,$$

where $\Sigma_1^+(t)$, $\Sigma_1^-(t)$ are defined by (3.36) and $\Sigma_2(t)$ is defined by (3.34). By (4.90), we get $\Sigma_1^{\pm}(t) < \infty$ in Examples 4.4 and 4.5. Hence, the orthogonality condition $(\mu^3_{(b,a)})^{L_t} \perp \mu^3_{(b,a)}$ for $t \in \pm O(3) \setminus \{e\}$ is equivalent to $\Sigma_2(t) = \infty$. Further, to prove irreducibility in Examples 4.4 and 4.5, we should show that $\Sigma_2(t) = \infty$ for all $t \in \pm O(3) \setminus \{e\}$ implies

$$\|C_1 Y_1 + C_2 Y_2 + C_3 Y_3\|^2 = \infty \quad \text{for all } (C_1, C_2, C_3) \in \mathbb{R}^3 \setminus \{0\}.$$

Lemma 4.18. (1) *The representations corresponding to the measures in Example 4.4 (a) and (b) are reducible.*
(2) *The representations corresponding to the measures in Example 4.5 are irreducible.*

Proof. To prove Part (1) of lemma, by Remark 4.15 and (4.90), we should find for the measure in Example 4.4 an element $t \in \pm O(3) \setminus \{e\}$ such that $\Sigma_2(t) < \infty$. This will imply $(\mu^3_{(b,a)})^{L_t} \sim \mu^3_{(b,a)}$ and hence *reducibility*. Finally, it is sufficient to find $t \in \pm O(3) \setminus \{e\}$ such that

$$t - 1 = \begin{pmatrix} \lambda_1 C_1 & \lambda_1 C_2 & \lambda_1 C_3 \\ \lambda_2 C_1 & \lambda_2 C_2 & \lambda_2 C_3 \\ \lambda_3 C_1 & \lambda_3 C_2 & \lambda_3 C_3 \end{pmatrix}, \tag{4.91}$$

where $(C_1, C_2, C_3) = (1, 1, -1)$, in Part (a), or for an arbitrary $(C_1, C_2, C_3) \in \mathbb{R}^3 \setminus \{0\}$ in Part (b). Such an element exists due to

Lemma 4.19. For such an element t, we get respectively for Cases (a) and (b) and Example 4.5 (see (3.34)),

$$\Sigma_2(t^{-1}) = \sum_{n\in\mathbb{Z}}(b_{1n}\lambda_1^2 + b_{2n}\lambda_2^2 + b_{3n}\lambda_3^2)(a_{1n} + a_{2n} - a_{3n})^2 = 0,$$

$$\Sigma_2(t^{-1}) = \sum_{n\in\mathbb{Z}}(b_{1n}\lambda_1^2 + b_{2n}\lambda_2^2 + b_{3n}\lambda_3^2)(C_1 a_{1n} + C_2 a_{2n} + C_3 a_{3n})^2 < \infty,$$

$$\Sigma_2(t^{-1}) = \sum_{n\in\mathbb{Z}}(b_{1n}\lambda_1^2 + b_{2n}\lambda_2^2 + b_{3n}\lambda_3^2)(C_1 a_{1n} + C_2 a_{2n} + C_3 a_{3n})^2 = \infty.$$

$$(4.92)$$

Note that the measure in Example 4.4 does not satisfy the orthogonality conditions.

Irreducibility. To prove Part (2) of the lemma, in Example 4.5, we cannot approximate x_{rn} by Lemma 5.1 since all the expressions

$$\Delta(Y_1^{(1)}, Y_2^{(1)}, Y_3^{(1)}), \quad \Delta(Y_2^{(2)}, Y_3^{(2)}, Y_1^{(2)}), \quad \Delta(Y_3^{(3)}, Y_1^{(3)}, Y_2^{(3)})$$

are bounded. To approximate D_{rn} using Lemma 5.2, we should estimate the following expressions:

$$\Delta(Y_1, Y_2, Y_3), \quad \Delta(Y_2, Y_3, Y_1), \quad \Delta(Y_3, Y_1, Y_2).$$

By Lemma 6.4 for $m = 2$, all these expressions are infinite if for all $(C_1, C_2, C_3) \in \mathbb{R}^3 \setminus \{0\}$, it holds that

$$\nu(C_1, C_2, C_3) := \|C_1 Y_1 + C_2 Y_2 + C_3 Y_3\|^2$$

$$= \sum_{n\in\mathbb{Z}} \frac{(C_1 a_{1n} + C_2 a_{2n} + C_3 a_{3n})^2}{\frac{1}{2b_{1n}} + \frac{1}{2b_{2n}} + \frac{1}{2b_{3n}}} = \infty.$$

In Example 4.5, we have

$$\nu(C_1, C_2, C_3) = \|C_1 Y_1 + C_2 Y_2 + C_3 Y_3\|^2$$

$$\sim \sum_{k\in\mathbb{Z}} b_{1n}(C_1 a_{1k} + C_2 a_{2k} + C_3 a_{3k})^2$$

$$\sim \sum_{n\in\mathbb{Z}}(b_{1n}\lambda_1^2 + b_{2n}\lambda_2^2 + b_{3n}\lambda_3^2)(C_1 a_{1n} + C_2 a_{2n} + C_3 a_{3n})^2$$

$$= \Sigma_2(t^{-1}) = \infty. \qquad \square$$

Lemma 4.19. *For an arbitrary $(C_1, C_2, C_3) \in \mathbb{R}^3 \setminus \{0\}$ and an arbitrary $D_3(s) = \mathrm{diag}(s_1, s_2, s_3)$ with $(s_1, s_2, s_3) \in (\mathbb{R}_+)^3$, there exists a unique element $t \in \pm O(3) \setminus \{e\}$ and $(\lambda_1, \lambda_2, \lambda_3) \in \mathbb{R}^3 \setminus \{0\}$ such that*

$$
D_3(s) t D_3^{-1}(s) - \mathrm{I} = \begin{pmatrix} \lambda_1 C_1 & \lambda_1 C_2 & \lambda_1 C_3 \\ \lambda_2 C_1 & \lambda_2 C_2 & \lambda_2 C_3 \\ \lambda_3 C_1 & \lambda_3 C_2 & \lambda_3 C_3 \end{pmatrix}
$$

$$
= \begin{pmatrix} \lambda_1 & 0 & 0 \\ 0 & \lambda_2 & 0 \\ 0 & 0 & \lambda_3 \end{pmatrix} \begin{pmatrix} 1 & 1 & 1 \\ 1 & 1 & 1 \\ 1 & 1 & 1 \end{pmatrix} \begin{pmatrix} C_1 & 0 & 0 \\ 0 & C_2 & 0 \\ 0 & 0 & C_3 \end{pmatrix} . \quad (4.93)
$$

Proof. By (4.93), we get

$$
\begin{pmatrix} e_1 \\ e_2 \\ e_3 \end{pmatrix} := t = \begin{pmatrix} t_{11}\ t_{12}\ t_{13} \\ t_{21}\ t_{22}\ t_{23} \\ t_{31}\ t_{32}\ t_{33} \end{pmatrix} = \begin{pmatrix} C_1\lambda_1 + 1 & \frac{s_2}{s_1} C_2\lambda_1 & \frac{s_3}{s_1} C_3\lambda_1 \\ \frac{s_1}{s_2} C_1\lambda_2 & C_2\lambda_2 + 1 & \frac{s_3}{s_2} C_3\lambda_2 \\ \frac{s_1}{s_3} C_1\lambda_3 & \frac{s_2}{s_3} C_2\lambda_3 & C_3\lambda_3 + 1 \end{pmatrix},
$$

$$
(4.94)
$$

$$
\text{where} \quad \|e_k\|^2 = 1 \quad \text{and} \quad e_k \perp e_r, \quad 1 \le k < r \le 3. \quad (4.95)
$$

By (4.94) and the first relations in (4.95), we get

$$
\lambda_k = -\frac{2 s_k^2 C_k}{s_1^2 C_1^2 + s_2^2 C_2^2 + s_3^2 C_3^2}, \quad 1 \le k \le 3. \quad (4.96)
$$

Then, the matrix elements $t = (t_{kr})_{k,r=1}^3$ are defined by (4.94). To verify $e_k \perp e_r$, we need to show that

$$
(e_1, e_2) = \frac{(s_1^2 C_1^2 + s_2^2 C_2^2 + s_3^2 C_3^2)\lambda_1\lambda_2}{s_1 s_2} + \frac{s_1^2 C_1\lambda_2 + s_2^2 C_2\lambda_1}{s_1 s_2} = 0,
$$

$$
(e_1, e_3) = \frac{(s_1^2 C_1^2 + s_2^2 C_2^2 + s_3^2 C_3^2)\lambda_1\lambda_3}{s_1 s_3} + \frac{s_1^2 C_1\lambda_3 + s_3^2 C_3\lambda_1}{s_1 s_3} = 0,
$$

$$
(e_2, e_3) = \frac{(s_1^2 C_1^2 + s_2^2 C_2^2 + s_3^2 C_3^2)\lambda_2\lambda_3}{s_2 s_3} + \frac{s_2^2 C_2\lambda_3 + s_3^2 C_3\lambda_2}{s_2 s_3} = 0.
$$

Indeed, for example, for (e_1, e_2), we have

$$(e_1, e_2) = \frac{(s_1^2 C_1^2 + s_2^2 C_2^2 + s_3^2 C_3^2)\lambda_1\lambda_2}{s_1 s_2} + \frac{s_1^2 C_1\lambda_2 + s_2^2 C_2\lambda_1}{s_1 s_2}$$

$$= \frac{1}{s_1 s_2(s_1^2 C_1^2 + s_2^2 C_2^2 + s_3^2 C_3^2)}(4s_1^2 s_2^2 - (2s_1^2 s_2^2 + 2s_1^2 s_2^2))C_1 C_2 = 0.$$

The proofs of $e_1 \perp e_3$ and $e_2 \perp e_3$ are similar. $\qquad\square$

Similarly, for any $m \geq 2$, we can prove the following lemma.

Lemma 4.20. *For an arbitrary $(C_k)_{k=1}^m \in \mathbb{R}^m \setminus \{0\}$ and $D_m(s) = \mathrm{diag}(s_k)_{k=1}^m$ with $s_k \in \mathbb{R}_+$, $1 \leq k \leq m$, there exists a unique element $t \in \pm O(m) \setminus \{e\}$ and $(\lambda_k)_{k=1}^m \in \mathbb{R}^m \setminus \{0\}$ such that*

$$D_m(s)tD_m^{-1}(s) - \mathrm{I} = \begin{pmatrix} \lambda_1 C_1 & \lambda_1 C_2 & \dots & \lambda_1 C_m \\ \lambda_2 C_1 & \lambda_2 C_2 & \dots & \lambda_2 C_m \\ & & \dots & \\ \lambda_m C_1 & \lambda_m C_2 & \dots & \lambda_m C_m \end{pmatrix}. \qquad (4.97)$$

The formulas for the corresponding λ_k are as follows:

$$\lambda_k = -2s_k^2 C_k \left(\sum_{r=1}^m s_r^2 C_r^2\right)^{-1}, \quad 1 \leq k \leq m. \qquad (4.98)$$

4.5.2   *Case $\Sigma_{123}(s) = (0, 1, 1)$*

We have for some $s_1 \in \mathbb{R}_+$ and all $(s_2, s_3) \in (\mathbb{R}_+)^2$,

$$\Sigma_{23}(s_2) = \infty, \quad \Sigma_{13}(s_3) = \infty.$$

Remark 4.16. Since $\Sigma_{12}(s_1) < \infty$, by (4.17) and Lemma 6.8, we can suppose

$$b = (b_{1n}, b_{2n}, b_{3n})_{n\in\mathbb{Z}} = (1, s_1^4(1 + c_n), b_{3n})_{n\in\mathbb{Z}}, \quad \sum_n c_n^2 < \infty.$$

Therefore, we can take $b = (1, 1, b_{3n})_{n\in\mathbb{Z}}$, $s = 1$, $c_n \equiv 0$.

Since $\Sigma_{13}(s) = \sum_{n\in\mathbb{Z}}\left(\frac{s^2}{\sqrt{b_{3n}}} - \frac{\sqrt{b_{3n}}}{s^2}\right)^2 = \infty$, we have, as in (4.48), three cases:

$$\lim_n b_{3n} = \begin{cases} (a)\ \infty, \\ (b)\ b > 0 \quad \text{with}\ \sum_n b_n^2 = \infty, \\ (c)\ 0, \end{cases} \tag{4.99}$$

where $b_{3n} = b(1 + b_n)$ with $\lim_n b_n = 0$ in Case (b). Note that the condition $S_3(3) = \infty$ implies $\sum_n b_{3n}^2 = \infty$. Indeed, by (4.6), we have

$$S_1(3) = S_2(3) = \sum_{n\in\mathbb{Z}} \frac{1}{1 + 2b_{3n}} = \infty,$$

$$\infty = S_3(3) = \sum_{n\in\mathbb{Z}} \frac{b_{3n}^2}{1 + 2b_{3n}} \overset{(3.17)}{\sim} \sum_{n\in\mathbb{Z}} b_{3n}^2. \tag{4.100}$$

By (4.4), we have

$$\|Y_r^{(r)}\|^2 = \sum_{k\in\mathbb{Z}} \frac{b_{rk}^2}{b_{rk}^2 + 2(b_{1n}b_{2n} + b_{1n}b_{3n} + b_{2n}b_{3n})},$$

$$\|Y_r^{(s)}\|^2 = \sum_{k\in\mathbb{Z}} \frac{b_{rk}^2}{b_{sk}^2 + 2(b_{1n}b_{2n} + b_{1n}b_{3n} + b_{2n}b_{3n})}, \quad s \neq r.$$

Let us denote

$$\begin{pmatrix} Y_{1n}^{(1)} & Y_{2n}^{(1)} & Y_{3n}^{(1)} \\ Y_{1n}^{(2)} & Y_{2n}^{(2)} & Y_{3n}^{(2)} \\ Y_{1n}^{(3)} & Y_{2n}^{(3)} & Y_{3n}^{(3)} \end{pmatrix} = \begin{pmatrix} \frac{1}{\sqrt{3+4b_{3n}}} & \frac{1}{\sqrt{3+4b_{3n}}} & \frac{b_{3n}}{\sqrt{3+4b_{3n}}} \\ \frac{1}{\sqrt{3+4b_{3n}}} & \frac{1}{\sqrt{3+4b_{3n}}} & \frac{b_{3n}}{\sqrt{3+4b_{3n}}} \\ \frac{1}{\sqrt{b_{3n}^2+4b_{3n}+2}} & \frac{1}{\sqrt{b_{3n}^2+4b_{3n}+2}} & \frac{b_{3n}}{\sqrt{b_{3n}^2+4b_{3n}+2}} \end{pmatrix}. \tag{4.101}$$

We have $\Delta(Y_1^{(1)}, Y_2^{(1)}, Y_3^{(1)}) = \Delta(Y_2^{(2)}, Y_3^{(2)}, Y_1^{(2)}) < \infty$. Indeed, since $Y_1^{(2)} = Y_2^{(2)}$, we get, for example,

$$\Delta(Y_2^{(2)}, Y_3^{(2)}, Y_1^{(2)}) = \frac{\Gamma(Y_2^{(2)}) + \Gamma(Y_2^{(2)}, Y_3^{(2)})}{1 + \Gamma(Y_3^{(2)}) + \Gamma(Y_1^{(2)}) + \Gamma(Y_3^{(2)}, Y_1^{(2)})} < 1.$$

Lemma 4.21. *In Cases (a), (b) and (c), given by (4.99), we have*

$$\Delta(Y_3^{(3)}, Y_1^{(3)}, Y_2^{(3)}) = \infty. \tag{4.102}$$

Proof. In all these cases, we have $Y_1^{(3)} = Y_2^{(3)}$. Hence, $\Gamma(Y_3^{(3)}, Y_1^{(3)}, Y_2^{(3)}) = 0$ and $\Gamma(Y_1^{(3)}, Y_2^{(3)}) = 0$. Therefore, by (3.15),

$$
\begin{aligned}
\Delta(Y_3^{(3)}, Y_1^{(3)}, Y_2^{(3)}) &= \frac{\Gamma(Y_3^{(3)}) + \Gamma(Y_3^{(3)}, Y_1^{(3)}) + \Gamma(Y_3^{(3)}, Y_2^{(3)})}{1 + \Gamma(Y_1^{(3)}) + \Gamma(Y_2^{(3)})} \\
&= \frac{\Gamma(Y_3^{(3)}) + 2\Gamma(Y_3^{(3)}, Y_1^{(3)})}{1 + 2\Gamma(Y_1^{(3)})} \sim \Delta(Y_3^{(3)}, Y_1^{(3)}).
\end{aligned}
\tag{4.103}
$$

We have two cases:

(a(1)) when $\|Y_1^{(3)}\| < \infty$ and (a(2)) when $\|Y_1^{(3)}\| = \infty$. In Case (a(1)), we have $\Delta(Y_3^{(3)}, Y_1^{(3)}) \sim \Gamma(Y_3^{(3)}) = \infty$. Therefore, (4.102) holds. In Case (a(2)), we should verify that

$$\|C_1 Y_1^{(3)} + C_3 Y_3^{(3)}\|^2 = \infty \quad \text{for all} \quad (C_1, C_3) \in \mathbb{R}^2 \setminus \{0\}. \tag{4.104}$$

Then, this will imply (4.102). We have

$$\|C_1 Y_1^{(3)} + C_3 Y_3^{(3)}\|^2 = \sum_{n \in \mathbb{Z}} \frac{(C_1 + C_3 b_{3n})^2}{b_{3n}^2 + 4b_{3n} + 2} =: \sum_{n \in \mathbb{Z}} g_n.$$

If $C_1 = 0$ or $C_3 = 0$, the later expression is divergent since $Y_1^{(3)} = Y_3^{(3)} = \infty$. Let $C_1 C_3 \neq 0$. In this case, $\lim_n g_n = C_3^2 > 0$ since $\lim_n b_{3n} = \infty$ from Case (a). Therefore, $\sum_{n \in \mathbb{Z}} g_n = \infty$. By Lemma 6.4 for $m = 1$, it implies that $\Delta(Y_3^{(3)}, Y_1^{(3)}) = \infty$; therefore, (4.102). In Case (b), we have by (4.103),

$$\Delta(Y_3^{(3)}, Y_1^{(3)}, Y_2^{(3)}) = \Delta(Y_3^{(3)}, Y_1^{(3)}).$$

To prove that $\Delta(Y_3^{(3)}, Y_1^{(3)}) = \infty$ using Lemma 6.4 for $m = 1$, we should verify (4.104). We have $\|Y_3^{(3)}\|^2 = \infty$ since $S = (0, 1, 1)$. By (4.101),

$$\|Y_1^{(3)}\|^2 = \sum_{n \in \mathbb{Z}} \frac{1}{b_{3n}^2 + 4b_{3n} + 2} \sim \sum_{n \in \mathbb{Z}} \frac{1}{b^2 + 4b + 2} = \infty.$$

The expression $\|C_1 Y_1^{(3)} + C_3 Y_3^{(3)}\|^2$ can be finite only for $(C_1, C_3) = \lambda(b, -1)$. Taking $\lambda = 1$, we get in Case (b),

$$
\begin{aligned}
\|C_1 Y_1^{(3)} + C_3 Y_3^{(3)}\|^2 &= \sum_{n \in \mathbb{Z}} \frac{(b - b_{3n})^2}{b_{3n}^2 + 4b_{3n} + 2} \\
&= \sum_{n \in \mathbb{Z}} \frac{b^2 b_n^2}{b^2(1 + b_n)^2 + 4b(1 + b_n) + 2} \\
&\overset{(3.19)}{\sim} \sum_{n \in \mathbb{Z}} \frac{b_n^2}{(4b + 2b^2)b_n + b^2 + 4b + 2} \\
&\overset{(3.17)}{\sim} \sum_{n \in \mathbb{Z}} b_n^2 = \infty.
\end{aligned}
$$

In Case (c), we have by (4.103),

$$
\Delta(Y_3^{(3)}, Y_1^{(3)}, Y_2^{(3)}) \sim \Delta(Y_3^{(3)}, Y_1^{(3)}).
$$

To prove that $\Delta(Y_3^{(3)}, Y_1^{(3)}) = \infty$ using Lemma 6.4 for $m = 1$, we should verify (4.104). Again, we have $\|Y_3^{(3)}\|^2 = \infty$ since $S = (0, 1, 1)$. Because of $\lim_n b_{3n} = 0$, we have by (4.101),

$$
\|Y_1^{(3)}\|^2 = \sum_{n \in \mathbb{Z}} \frac{1}{b_{3n}^2 + 4b_{3n} + 2} \sim \sum_{n \in \mathbb{Z}} \frac{1}{2} = \infty.
$$

Let $C_1 C_3 \neq 0$. Then, since $\lim_n b_{3n} = 0$, we get

$$
\begin{aligned}
\|C_1 Y_1^{(3)} + C_3 Y_3^{(3)}\|^2 &= \sum_{n \in \mathbb{Z}} \frac{(C_1 + C_3 b_{3n})^2}{b_{3n}^2 + 4b_{3n} + 2} \\
&= \sum_{n \in \mathbb{Z}} \frac{C_3^2 (b_{3n} + C_1 C_3^{-1})^2}{b_{3n}^2 + 4b_{3n} + 2} = \infty. \qquad \square
\end{aligned}
$$

By Lemma 4.21, we can approximate x_{3n}. By (4.5), we have

$$
\|Y_1\|^2 = \sum_{n \in \mathbb{Z}} \frac{a_{1n}^2}{\frac{1}{2b_{1n}} + \frac{1}{2b_{2n}} + \frac{1}{2b_{3n}}} = \sum_{k \in \mathbb{Z}} \frac{a_{1n}^2}{1 + \frac{1}{2b_{3n}}} = \sum_{k \in \mathbb{Z}} \frac{2b_{3n} a_{1n}^2}{1 + 2b_{3n}},
$$

$$\|Y_2\|^2 = \sum_{n\in\mathbb{Z}} \frac{a_{2n}^2}{1 + \frac{1}{2b_{3n}}} = \sum_{k\in\mathbb{Z}} \frac{2b_{3n}a_{2n}^2}{1 + 2b_{3n}},$$

$$\|Y_3\|^2 = \sum_{n\in\mathbb{Z}} \frac{a_{3n}^2}{1 + \frac{1}{2b_{3n}}} = \sum_{k\in\mathbb{Z}} \frac{2b_{3n}a_{3n}^2}{1 + 2b_{3n}}.$$

Therefore, in Cases (a) and (b), we have

$$\|Y_1\|^2 \sim \sum_{k\in\mathbb{Z}} a_{1n}^2, \quad \|Y_2\|^2 \sim \sum_{k\in\mathbb{Z}} a_{2n}^2, \quad \|Y_3\|^2 \sim \sum_{k\in\mathbb{Z}} a_{3n}^2.$$

In Case (c), we get

$$\|Y_1\|^2 \sim \sum_{k\in\mathbb{Z}} b_{3n}a_{1n}^2, \quad \|Y_2\|^2 \sim \sum_{k\in\mathbb{Z}} b_{3n}a_{2n}^2, \quad \|Y_3\|^2 \sim \sum_{k\in\mathbb{Z}} b_{3n}a_{3n}^2.$$

Since in Cases (a) and (b)

$$\|Y_1\|^2 \sim \sum_{n\in\mathbb{Z}} a_{1n}^2 = \sum_{n\in\mathbb{Z}} b_{1n}a_{1n}^2 \sim S_{11}^L(\mu) = \infty,$$

$$\|Y_2\|^2 \sim \sum_{n\in\mathbb{Z}} a_{2n}^2 = \sum_{n\in\mathbb{Z}} b_{2n}a_{2n}^2 = S_{22}^L(\mu) = \infty,$$

we have two possibilities for $y_{23} := (y_2, y_3) \in \{0,1\}^2$, referring to Section 4.4.4:

	(1.1)	(1.3)
y_1	1	1
y_2	1	1
y_3	0	1

In Case (c), we have

$$\|Y_3\|^2 \sim \sum_{n\in\mathbb{Z}} b_{3n}a_{3n}^2 \sim S_{33}^L(\mu) = \infty.$$

Therefore, we have four possibilities for $y_{12} := (y_1, y_2) \in \{0,1\}^2$, referring to (4.52):

	(1.0)	(1.1)	(1.2)	(1.3)
y_1	0	1	0	1
y_2	0	0	1	1
y_3	1	1	1	1

Further, in Cases (a) and (b), we have four possibilities: (1.1.1), (1.3.1) and (1.1.0), (1.3.0); see Remark 4.5. In Case (1.1.1), we can approximate D_{1n}, D_{2n}, and in Case (1.3.1), we can approximate all D_{rn}, $1 \leq r \leq 3$. In these cases, the proof is completed since we get, respectively, $D_{1n}, D_{2n}, x_{3n} \, \eta \, \mathfrak{A}^3$. The subcases (1.1.0) and (1.3.0) of Cases (a) and (b) are considered below.

In the subcase (1.0) of Case (c), we can approximate D_{3n} using Lemma 5.2 since $\Delta(Y_3, Y_2, Y_1) \sim \|Y_3\|^2 = \infty$; therefore, we have $D_{3n}, x_{3n} \, \eta \, \mathfrak{A}^3$, and the proof is completed. Further, in Case (c), we have six subcases: (1.1.1), (1.2.1), (1.3.1) and (1.1.0), (1.2.0), (1.3.0), according to whether the corresponding expressions are divergent; see Remark 4.5. We can approximate in the three first subcases: D_{1n} and D_{3n} in Case (1.1.1), D_{2n} and D_{3n} in Case (1.1.2) and all D_{1n}, D_{2n}, D_{3n} in (1.1.3). The proof of irreducibility is completed in these subcases because we have, respectively, $D_{1n}, D_{3n}, x_{3n} \, \eta \, \mathfrak{A}^3$, $D_{2n}, D_{3n}, x_{3n} \, \eta \, \mathfrak{A}^3$ and $D_{1n}, D_{2n}, D_{3n}, x_{3n} \, \eta \, \mathfrak{A}^3$.

If the opposite holds, in Cases (a), (b) and (c), i.e., we are in Cases (1.1.0), (1.2.0) and (1.3.0), respectively, we try to approximate D_{3n} using Lemma 5.4. If one of the expressions $\Sigma_3(D, s)$ or $\Sigma_3^{\vee}(D, s)$ is divergent, we can approximate D_{3k}, and the proof is completed since we have $x_{3n}, D_{3n} \, \eta \, \mathfrak{A}^3$. Let us suppose, as in Remark 4.3, that for every sequence $s = (s_k)_{k \in \mathbb{Z}}$,

$$\Sigma_3(D, s) + \Sigma_3^{\vee}(D, s) < \infty.$$

Then, in particular, we have for $s^{(3)} = (s_k)_{k \in \mathbb{Z}}$, with $\frac{s_k^2}{b_{3k}} \equiv 1$,

$$\infty \; > \; \Sigma_3(D, s^{(3)}) + \Sigma_3^{\vee}(D, s^{(3)}) \sim \Sigma_3(D) + \Sigma_3^{\vee}(D)$$

$$= \; \sum_k \frac{\frac{1}{2b_{3k}} + a_{3k}^2}{C_k + a_{1k}^2 + a_{2k}^2 + a_{3k}^2}$$

$$\overset{(3.19)}{\sim} \; \sum_k \frac{\frac{1}{2b_{3k}} + a_{3k}^2}{\frac{1}{2b_{1k}} + a_{1k}^2 + \frac{1}{2b_{2k}} + a_{2k}^2} = \sum_k \frac{\frac{1}{2b_{3k}} + a_{3k}^2}{1 + a_{1k}^2 + a_{2k}^2} =: \Sigma_3^{\vee,+}(D).$$

$$(4.105)$$

In Cases (a), (b) and (c), we have, respectively,

$$\Sigma_3^{\vee,+}(D) \sim \Sigma_3^+(D) = \sum_k \frac{2a_{3k}^2}{1 + 2a_{1k}^2 + 2a_{2k}^2},$$

$$\Sigma_3^{\vee,+}(D) = \sum_k \frac{\frac{1}{2b_{3k}} + a_{3k}^2}{1 + a_{1k}^2 + a_{2k}^2}.$$

In particular, in Case (c), we have by (4.105),

$$\infty > \sum_k \frac{\frac{1}{2b_{3k}} + a_{3k}^2}{1 + a_{1k}^2 + a_{2k}^2} > \sum_k \frac{a_{3k}^2}{1 + a_{1k}^2 + a_{2k}^2} \sim \Sigma_3^+(D). \qquad (4.106)$$

The subcase (1.1.0) of Case (a), where $\|Y_3\|^2 < \infty$, cannot occur because the conditions $\Sigma_{12}(s_1) < \infty$ and $\nu_{12}(C_1, C_2) < \infty$, defined by (4.56), contradict the orthogonality condition for the matrix $\tau_{12}(\phi, s)$ defined by (4.66). Indeed, by Remark 3.8 (where, instead of $\mu_{(b,a)}^2$, we can write $\mu_{(b,a)}^3$),

$$\left(\mu_{(b,a)}^3\right)^{L_{\tau_{12}(\phi,s)}} \perp \mu_{(b,a)}^3 \Leftrightarrow \Sigma_{12}(s) + \Sigma_{12}(C_1, C_2) = \infty,$$

where Σ_{12} is defined by (4.67) and $\Sigma_{12}(C_1, C_2)$ is defined by (4.68):

$$\Sigma_{12}(C_1, C_2) := \sum_{n \in \mathbb{Z}} (C_1^2 b_{1n} + C_2^2 b_{2n})(C_1 a_{1n} + C_2 a_{2n})^2 \sim \nu_{12}(C_1, C_2),$$

$$\infty > \Sigma_{12}(s) + \nu_{12}(C_1, C_2) \sim \Sigma_{12}(s) + \Sigma_{12}(C_1, C_2) = \infty, \quad (4.107)$$

which is a contradiction. In the subcase (1.3.0) of Cases (a) and (b), we get $\Sigma_3^+(D) = \infty$ by Lemma 4.12, a contradiction with (4.105). Hence, $D_{3n}\, \eta\, \mathfrak{A}^3$. In the subcases (1.1.0) and (1.2.0) of Case (c), we have, respectively, $\|Y_2\|^2 < \infty$ and $\|Y_1\|^2 < \infty$. Hence,

$$\Sigma_3^+(D) \sim \sum_k \frac{a_{3k}^2}{1 + a_{1k}^2} = \infty, \quad \Sigma_3^+(D) \sim \sum_k \frac{a_{3k}^2}{1 + a_{2k}^2} = \infty.$$

By Lemma 4.11, we once again face a contradiction with (4.105). Hence, $D_{3n}\, \eta\, \mathfrak{A}^3$. In the subcase (1.3.0) of Case (c), we get

$$\Sigma_3^+(D) = \sum_k \frac{a_{3k}^2}{1 + a_{1k}^2 + a_{2k}^2} = \infty$$

by Lemma 4.12, which is contradictory with (4.105). Hence, $D_{3n}\, \eta\, \mathfrak{A}^3$.

4.5.3 *Case* $\Sigma_{123}(s) = (1, 1, 1)$

We have for all $s = (s_{12}, s_{23}, s_{13}) \in \mathbb{R}_+^3 \setminus \{0\}$,

$$\Sigma_{12}(s_{12}) = \infty, \quad \Sigma_{23}(s_{23}) = \infty, \quad \Sigma_{13}(s_{13}) = \infty, \qquad (4.108)$$

$$b = (b_{1n}, b_{2n}, b_{3n})_{n \in \mathbb{Z}} \overset{(4.17)}{=} (1, d_{2n}, d_{3n})_{n \in \mathbb{Z}}.$$

Recall that, referring to (4.31), we denote $D_n := d_{2n}^{-1} + d_{3n}^{-1} + 1$ and $d_n = \frac{d_{3n}}{d_{3n}}$. Set

$$\begin{pmatrix} Y_{1n}^{(1)} & Y_{2n}^{(1)} & Y_{3n}^{(1)} \\[4pt] Y_{1n}^{(2)} & Y_{2n}^{(2)} & Y_{3n}^{(2)} \\[4pt] Y_{1n}^{(3)} & Y_{2n}^{(3)} & Y_{3n}^{(3)} \end{pmatrix}$$

$$= \begin{pmatrix} \dfrac{1}{\sqrt{1+2D_n d_{2n} d_{3n}}} & \dfrac{d_{2n}}{\sqrt{1+2D_n d_{2n} d_{3n}}} & \dfrac{d_{3n}}{\sqrt{1+2D_n d_{2n} d_{3n}}} \\[12pt] \dfrac{1}{\sqrt{d_{2n}^2+2D_n d_{2n} d_{3n}}} & \dfrac{d_{2n}}{\sqrt{d_{2n}^2+2D_n d_{2n} d_{3n}}} & \dfrac{d_{3n}}{\sqrt{d_{2n}^2+2D_n d_{2n} d_{3n}}} \\[12pt] \dfrac{1}{\sqrt{d_{3n}^2+2D_n d_{2n} d_{3n}}} & \dfrac{d_{2n}}{\sqrt{d_{3n}^2+2D_n d_{2n} d_{3n}}} & \dfrac{d_{3n}}{\sqrt{d_{3n}^2+2D_n d_{2n} d_{3n}}} \end{pmatrix}.$$

$$(4.109)$$

Remark 4.17. For (r, s) such that $1 \leq r < s \leq 3$, the following equivalence holds:

$$\Sigma_{rs}(s_{rs}) < \infty \Leftrightarrow \sum_{n \in \mathbb{Z}} c_{rs,n}^2 < \infty \Leftrightarrow \sum_{n \in \mathbb{Z}} c_{sr,n}^2 < \infty, \quad \text{where}$$

$$(4.110)$$

$$\frac{b_{rn}}{b_{sn}} =: s_{rs}^{-4}(1 + c_{rs,n}), \quad \frac{b_{sn}}{b_{rn}} = s_{rs}^{4}(1 + c_{sr,n}), \quad \lim_{n} \frac{b_{rn}}{b_{sn}} \in (0, \infty).$$

$$(4.111)$$

Proof. By Lemma 6.8, we have

$$\Sigma_{rs}(s_{rs}) = \sum_{n \in \mathbb{Z}} \left(s_{rs}^{2} \sqrt{\frac{b_{rn}}{b_{sn}}} - s_{rs}^{-2} \sqrt{\frac{b_{sn}}{b_{rn}}} \right)^{2} = \sum_{n \in \mathbb{Z}} \frac{c_{rs,n}^{2}}{1 + c_{rs,n}} \sim \sum_{n \in \mathbb{Z}} c_{rs,n}^{2},$$

$$\Sigma_{sr}(s_{rs}^{-1}) = \sum_{n \in \mathbb{Z}} \left(s_{rs}^{-2} \sqrt{\frac{b_{sn}}{b_{rn}}} - s_{rs}^{2} \sqrt{\frac{b_{rn}}{b_{sn}}} \right)^{2} = \sum_{n \in \mathbb{Z}} \frac{c_{sr,n}^{2}}{1 + c_{sr,n}} \sim \sum_{n \in \mathbb{Z}} c_{sr,n}^{2}.$$

Note also that

$$1 = \frac{b_{rn}}{b_{sn}} \frac{b_{sn}}{b_{rn}} = (1 + c_{rs,n})(1 + c_{sr,n}). \qquad \square$$

$$(4.112)$$

By Remark 4.17, the condition $\Sigma_{rs}(s_{rs}) = \infty$ holds in the following cases:

$$l_{sr} := \lim_{n} \frac{b_{sn}}{b_{rn}} = \begin{cases} (a) \ \infty, \\ (b) \ s_{rs}^{4} > 0 \quad \text{with} \quad \sum_{n \in \mathbb{Z}} c_{sr,n}^{2} = \infty, \\ (c) \ 0, \\ (d) \ \lim \qquad \text{does not exist.} \end{cases}$$

$$(4.113)$$

Remark 4.18. In Case (d), we can use the fact that some *subsequence* of $\left(\frac{b_{sn}}{b_{rn}} \right)_{n \in \mathbb{Z}}$ has the property (a), (b) or (c). We can avoid Case (c). Namely, if $l_{sr} = 0$ for some pair (r, s) with $1 \leq r < s \leq 3$, we can exchange the two lines (b_{sn}, a_{sn}) and (b_{rn}, a_{rn}) to obtain $l_{sr} = \infty$.

Formally, we have $3^{3} = \#(A)^{\#(B)}$ possibilities where $A = \{(21), (32), (31)\}$ and $B = \{(a), (b), (d)\}$. Since $l_{32}l_{21} = l_{31}$, we get only the following cases:

$$
\begin{array}{c|ccc}
e \setminus (rs) & (21) & (32) & (31) \\
(1) & b & b & b \\
(2) & a & a & a \\
(3) & a & b & a \\
(4) & b & a & b
\end{array}
$$

To be able to approximate x_{rn} for $1 \leq r \leq 3$, we should know when the following expressions are infinite:

$$\rho_r(C_1, C_2, C_3) = \|C_1 Y_1^{(r)} + C_2 Y_2^{(r)} + C_2 Y_3^{(r)}\|^2. \tag{4.114}$$

By (4.109), we have

$$\rho_r(C_1, C_2, C_3) =: \sum_n \frac{|C_1 + C_2 d_{2n} + C_3 d_{3n}|^2}{C_{rn}}, \tag{4.115}$$

where $\quad C_{1n} = 1 + 2D_n d_{2n} d_{3n}, \quad C_{2n} = d_{2n}^2 + 2D_n d_{2n} d_{3n},$

$$C_{3n} = d_{3n}^2 + 2D_n d_{2n} d_{3n}.$$

Consider **Case (1)=(bbb)**. We prove the following lemma.

Lemma 4.22. *Assume that (4.108) holds for all $s = (s_{12}, s_{23}, s_{13}) \in \left(\mathbb{R}_+\right)^3$. Then,*

$$\Delta(Y_3^{(3)}, Y_1^{(3)}, Y_2^{(3)}) = \Delta(Y_2^{(2)}, Y_3^{(2)}, Y_1^{(2)}) = \Delta(Y_1^{(1)}, Y_2^{(1)}, Y_3^{(1)}) = \infty. \tag{4.116}$$

Proof. For $1 \leq r < s \leq 3$, set

$$\frac{b_{sn}}{b_{rn}} = s_{rs}^4(1 + c_{sr,n}) \quad \text{with} \quad \sum_{n \in \mathbb{Z}} c_{sr,n}^2 = \infty, \quad \lim_{n \to \infty} c_{sr,n} = 0.$$

For $b_{1n} \equiv 1$, we have

$$b_{2n} = s_{12}^4(1 + c_{21,n}), \quad b_{3n} = s_{13}^4(1 + c_{31,n}),$$

$$\frac{b_{3n}}{b_{2n}} = \frac{s_{13}^4}{s_{12}^4}\frac{1 + c_{31,n}}{1 + c_{21,n}} = s_{23}^4(1 + c_{32,n}),$$

$$c_{32,n} = \frac{1 + c_{31,n}}{1 + c_{21,n}} - 1, \quad s_{23} = \frac{s_{13}}{s_{12}},$$

$$\sum_n c_{32,n}^2 = \sum_n \left(\frac{1 + c_{31,n}}{1 + c_{21,n}} - 1\right)^2$$

$$= \sum_n \left(\frac{c_{31,n} - c_{21,n}}{1 + c_{21,n}}\right)^2 \sim \sum_n (c_{21,n} - c_{31,n})^2 = \infty.$$

Finally, we get

$$\sum_n c_{21,n}^2 = \infty, \quad \sum_n c_{31,n}^2 = \infty, \quad \sum_n (c_{21,n} - c_{31,n})^2 = \infty. \qquad (4.117)$$

By (4.114) and (4.115), we get

$$\rho_r(C_1, C_2, C_3) = \|C_1 Y_1^{(r)} + C_2 Y_2^{(r)} + C_2 Y_3^{(r)}\|^2$$

$$= \sum_n \frac{|C_1 + C_2 d_{2n} + C_3 d_{3n}|^2}{C_{rn}}$$

$$= \sum_n \frac{|C_1 + C_2 s_{12}^4(1 + c_{21,n}) + C_3 s_{13}^4(1 + c_{31,n})|^2}{C_{rn}}.$$

The latter expression is divergent if $C_1 + C_2 s_{12}^4 + C_3 s_{13}^4 \neq 0$ since $\lim_{n\to\infty} c_{21,n} = \lim_{n\to\infty} c_{31,n} = 0$ and $A_1 \leq C_{rn} \leq A_2$. If $C_1 + C_2 s_{12}^4 + C_3 s_{13}^4 = 0$, we get

$$\rho_r(C_1, C_2, C_3) = \sum_{n\in\mathbb{Z}} \frac{|C_2 s_{12}^4 c_{21,n} + C_3 s_{13}^4 c_{31,n}|^2}{C_{rn}} =: \rho_r(C_2, C_3).$$

$$(4.118)$$

The latter expression is divergent due to the first two relations in (4.117) when (1) $C_2 C_3 > 0$, (2) $C_2 = 0$ and $C_3 \neq 0$, (3) $C_2 \neq 0$ and $C_3 = 0$. If

$C_2 C_3 < 0$, we have from the last relation in (4.117),

$$\sum_{n \in \mathbb{Z}} \frac{\left| C_2 s_{12}^4 c_{21,n} - C_3 s_{13}^4 c_{31,n} \right|^2}{C_{rn}} \sim \sum_{n \in \mathbb{Z}} \left| C_2 s_{12}^4 c_{21,n} - C_3 s_{13}^4 c_{31,n} \right|^2 = \infty$$

since $(s_{12}, s_{13}) = \frac{1}{s_1}(s_2, s_3) \in (\mathbb{R}^*)^2$ are arbitrary.

Consider **Case (2)=(aaa)**. Now, referring to (4.113), we have

$$l_{21} = \lim_n \frac{b_{2n}}{b_{1n}} = \infty, \quad l_{32} = \lim_n \frac{b_{3n}}{b_{2n}} = \infty, \quad \text{therefore,} \quad l_{31} = \lim_n \frac{b_{3n}}{b_{1n}} = \infty.$$
$$(4.119)$$

Since $b_{1n} \equiv 1$, we conclude that

$$l_{21} = \lim_n d_{2n} = \infty \quad \text{and} \quad l_{31} = \lim_n d_{3n} = \infty. \tag{4.120}$$

Therefore, we get for some $C > 0$ and all $n \in \mathbb{Z}$,

$$1 \leq D_n = 1 + \left(d_{2n} \right)^{-1} + \left(d_{3n} \right)^{-1} \leq C. \tag{4.121}$$

By (4.114) and (4.115), we obtain

$$\rho_r(C_1, C_2, C_3) = \| C_1 Y_1^{(r)} + C_2 Y_2^{(r)} + C_2 Y_3^{(r)} \|^2$$
$$= \sum_n \frac{\left| C_1 + C_2 d_{2n} + C_3 d_{3n} \right|^2}{C_{rn}}$$
$$\sim \sum_n \frac{\left| C_1 + C_2 d_{2n} + C_3 d_{3n} \right|^2}{C'_{rn}} =: \rho'_r(C_1, C_2, C_3),$$

where $C'_{rn} = 1 + 2d_{2n}d_{3n}, \quad C'_{rn} = d_{2n}^2 + 2d_{2n}d_{3n},$

$$C'_{rn} = d_{3n}^2 + 2d_{2n}d_{3n}.$$

We should know when $\rho'_r(C_1, C_2, C_3) = \infty$ for some $1 \leq r \leq 3$:

$$\sum_n \frac{\left| C_1 + C_2 d_{2n} + C_3 d_{3n} \right|^2}{1 + 2d_{2n}d_{3n}}, \quad \sum_n \frac{\left| C_1 + C_2 d_{2n} + C_3 d_{3n} \right|^2}{d_{2n}^2 + 2d_{2n}d_{3n}},$$

$$\sum_n \frac{\left| C_1 + C_2 d_{2n} + C_3 d_{3n} \right|^2}{d_{3n}^2 + 2d_{2n}d_{3n}}.$$

Writing as before $d_n = \frac{d_{3n}}{d_{2n}} =: l_n^{-1}$, we get

$$\rho_1'(C_1, C_2, C_3) = \sum_n \frac{|C_1 + C_2 d_{2n} + C_3 d_{3n}|^2}{1 + 2d_{2n}d_{3n}} = \sum_n \frac{|\frac{C_1}{d_{2n}} + C_2 + C_3 d_n|^2}{\frac{1}{d_{2n}^2} + 2d_n},$$

$$\sim \sum_n \frac{|C_2 + C_3 d_n|^2}{2d_n},$$

$$\rho_2'(C_1, C_2, C_3) = \sum_n \frac{|C_1 + C_2 d_{2n} + C_3 d_{3n}|^2}{d_{2n}^2 + 2d_{2n}d_{3n}} = \sum_n \frac{|\frac{C_1}{d_{2n}} + C_2 + C_3 d_n|^2}{1 + 2d_n}$$

$$\sim \sum_n \frac{|C_2 + C_3 d_n|^2}{1 + 2d_n},$$

$$\rho_3'(C_1, C_2, C_3) = \sum_n \frac{|C_1 + C_2 d_{2n} + C_3 d_{3n}|^2}{d_{3n}^2 + 2d_{2n}d_{3n}} = \sum_n \frac{|\frac{C_1}{d_{2n}} + C_2 + C_3 d_n|^2}{d_n^2 + 2d_n}$$

$$\sim \sum_n \frac{|C_2 + C_3 d_n|^2}{d_n^2 + 2d_n} = \sum_n \frac{|C_2 l_n + C_3|^2}{1 + 2l_n}.$$

By Lemma 4.9, we get when $C_2 C_3^{-1} > 0$:

$$\rho_1'(C_1, C_2, -C_3) \sim \sum_n \frac{|C_2 - C_3 d_n|^2}{2d_n} > \sum_n \frac{|C_2 - C_3 d_n|^2}{1 + 2d_n}$$

$$\sim \sum_n c_n^2 \sim \Sigma_{23}(s),$$

$$\rho_2'(C_1, C_2, -C_3) \sim \sum_n \frac{|C_2 - C_3 d_n|^2}{1 + 2d_n} \sim \sum_n c_n^2 \sim \Sigma_{23}(s),$$

$$\rho_3'(C_1, C_2, -C_3) \sim \sum_n \frac{|C_2 l_n - C_3|^2}{1 + 2l_n} \sim \sum_n e_n^2 \sim \Sigma_{23}(s), \qquad (4.122)$$

$$\text{where} \quad d_n = C_2 C_3^{-1}(1 + c_n), \quad l_n = C_3 C_2^{-1}(1 + e_n),$$

$$s^4 = C_2 C_3^{-1} > 0.$$

But $\Sigma_{23}(s) = \infty$ for all $s > 0$; therefore, for $C_2 C_3^{-1} > 0$, we have

$$\rho_r(C_1, C_2, -C_3) \sim \Sigma_{23}(s) = \infty. \tag{4.123}$$

If $C_2 C_3^{-1} > 0$, by (4.122), we get

$$\sum_n \frac{|C_2 + C_3 d_n|^2}{1 + 2d_n} > \sum_n \frac{|C_2 - C_3 d_n|^2}{1 + 2d_n} \sim \sum_n c_n^2 \sim \Sigma_{23}(s) = \infty,$$

$$\sum_n \frac{|C_2 + C_3 d_n|^2}{1 + 2d_n} \sum_n \frac{|C_2 - C_3 d_n|^2}{1 + 2d_n} \sim \sum_n c_n^2 \sim \Sigma_{23}(s) = \infty,$$

$$\sum_n \frac{|C_2 l_n + C_3|^2}{1 + 2l_n} > \sum_n \frac{|C_2 l_n - C_3|^2}{1 + 2l_n} \sim \sum_n e_n^2 \sim \Sigma_{23}(s) = \infty.$$

Therefore, $\rho_r(C_1, C_2, C_3) = \infty$ for every $(C_1, C_2, C_3) \in \mathbb{R}^3 \setminus \{0\}$.

Consider **Case (3)=(aba)**. Now, referring to (4.113), we have

$$l_{21} = \lim_n \frac{b_{2n}}{b_{1n}} = \infty, \quad l_{32} = \lim_n \frac{b_{3n}}{b_{2n}} < \infty,$$

$$\text{therefore,} \quad l_{31} = \lim_n \frac{b_{3n}}{b_{1n}} = \infty.$$

So, we have again, referring to (4.120),

$$l_{21} = \lim_n d_{2n} = \infty, \quad \text{and} \quad l_{31} = \lim_n d_{3n} = \infty.$$

This case is reduced to Case (2).

Consider **Case (4)=(baa)**. Now, referring to (4.113), we have

$$l_{21} = \lim_n d_{2n} < \infty, \quad \text{and} \quad l_{31} = \lim_n d_{3n} = \infty.$$

Hence, (4.121) holds as well, and we can use all the estimations from Case (1). $\qquad\square$

Remark 4.19. In Cases (1)–(4), by Lemmas 4.22 and 5.1, we can approximate x_{rn} for all $1 \leq r \leq 3$ and $n \in \mathbb{Z}$, and irreducibility is thus proved.

Chapter 5

Approximation

5.1   Approximation of x_{kn} by $A_{nk}A_{tk}$

For $m = 3$, consider three rows as follows:

$$\begin{pmatrix} \cdots & b_{11} & b_{12} & \cdots & b_{1n} & \cdots \\ \cdots & b_{21} & b_{22} & \cdots & b_{2n} & \cdots \\ \cdots & b_{31} & b_{32} & \cdots & b_{3n} & \cdots \end{pmatrix}.$$

Set $\lambda_k^{(r)} = (b_{1k} + b_{2k} + b_{3k})^2 - (b_{1k}^2 + b_{2k}^2 + b_{3k}^2 - b_{rk}^2)$,

$$r = 1, 2, 3, \ k \in \mathbb{Z}. \tag{5.1}$$

Denote by $Y_r^{(s)}$ the following vectors:

$$x_{rk}^{(s)} = b_{rk} / \sqrt{\lambda_k^{(s)}}, \ k \in \mathbb{Z}, \quad Y_r^{(s)} = (x_{rk}^{(s)})_{k \in \mathbb{Z}}. \tag{5.2}$$

Lemma 5.1. *For any $n, t \in \mathbb{Z}$ and $1 \leq r \leq 3$, one has*

$$x_{rn} x_{rt} 1 \in \langle A_{nk} A_{tk} 1 \mid k \in \mathbb{Z} \rangle \ \Leftrightarrow \ \Delta(Y_r^{(r)}, Y_s^{(r)}, Y_l^{(r)}) = \infty,$$

where $\{r, s, l\}$ is a cyclic permutation of $\{1, 2, 3\}$.

Proof. The proof of Lemma 5.1 for $r = 1$ is also based on Lemma 6.6 for $m = 2$. We study the case when $x_{1n}x_{1t}\mathbf{1} \in \langle A_{nk}A_{tk}\mathbf{1} \mid k \in \mathbb{Z}\rangle$. Since

$$A_{nk}A_{tk} = (x_{1n}D_{1k} + x_{2n}D_{2k} + x_{3n}D_{3k})(x_{1t}D_{1k} + x_{2t}D_{2k} + x_{3t}D_{3k})$$

$$= x_{1n}x_{1t}D_{1k}^2 + x_{2n}x_{2t}D_{2k}^2 + x_{3n}x_{3t}D_{3k}^2$$

$$+ (x_{1n}x_{2t} + x_{2n}x_{1t})D_{1k}D_{2k} + (x_{1n}x_{3t} + x_{3n}x_{1t})D_{1k}D_{3k}$$

$$+ (x_{2n}x_{3t} + x_{3n}x_{2t})D_{2k}D_{3k}$$

and $MD_{1k}^2\mathbf{1} = -\frac{b_{1k}}{2}$, we take $t = (t_k)$ as follows: $-\sum_{k=-m}^{m} t_k\frac{b_{1k}}{2} = (t, b') = 1$, where $t = (t_k)_{k=-m}^{m}$ and $b' = -(\frac{b_{1k}}{2})_{k=-m}^{m} \sim b = (b_{1k})_{k=-m}^{m}$. We have

$$\left\|\left[\sum_{k=-m}^{m} t_k A_{nk}A_{tk} - x_{1n}x_{1t}\right]\mathbf{1}\right\|^2$$

$$= \left\|\sum_{k=-m}^{m} t_k\left[x_{1n}x_{1t}\left(D_{1k}^2 + \frac{b_{1k}}{2}\right) + x_{2n}x_{2t}D_{2k}^2\right.\right.$$

$$+ x_{3n}x_{3t}D_{3k}^2 + (x_{1n}x_{2t} + x_{2n}x_{1t})D_{1k}D_{2k}$$

$$\left.\left.+ (x_{1n}x_{3t} + x_{3n}x_{1t})D_{1k}D_{3k} + (x_{2n}x_{3t} + x_{3n}x_{2t})D_{2k}D_{3k}\right]\mathbf{1}\right\|^2$$

$$= \sum_{-m \leq k,r \leq m} (f_k, f_r)t_k t_r =: (A_{2m+1}t, t),$$

where $A_{2m+1} = ((f_k, f_r))_{k,r=-m}^{m}$ and $f_k = \sum_{r=1}^{3} f_k^r + \sum_{1 \leq i < j \leq 3} f_k^{ij}$,

$$f_k^r = x_{rn}x_{rt}\left(D_{rk}^2 + \frac{b_{rk}}{2}\delta_{1r}\right)\mathbf{1}, \quad f_k^{ij} = (x_{in}x_{jt} + x_{jn}x_{it})D_{ik}D_{jk}\mathbf{1} \quad (5.3)$$

for $1 \leq r \leq 3$, $1 \leq i < j \leq 3$. Since $f_k^r \perp f_k^{ij}$, $f_k^{ij} \perp f_k^{i'j'}$ for different (ij), $(i'j')$, writing $c_{kn} = \|x_{kn}\|^2 = \frac{1}{2b_{kn}} + a_{kn}^2$, we get

$$(f_k, f_k) = \sum_{r=1}^{3} \|f_k^r\|^2 + \sum_{1 \leq i < j \leq 3} \|f_k^{ij}\|^2$$

$$= c_{1n}c_{1t}2\left(\frac{b_{1k}}{2}\right)^2 + c_{2n}c_{2t}3\left(\frac{b_{2k}}{2}\right)^2 + c_{3n}c_{3t}3$$

$$\times \left(\frac{b_{3k}}{2}\right)^2 + \left(c_{1n}c_{2t} + c_{1t}c_{2n} + 2a_{1n}a_{2t}a_{1t}a_{2n}\right)\frac{b_{1k}}{2}\frac{b_{2k}}{2}$$

$$+ \left(c_{1n}c_{3t} + c_{3t}c_{1n} + 2a_{1n}a_{3t}a_{3t}a_{1n}\right)$$

$$\times \frac{b_{1k}}{2}\frac{b_{3k}}{2} + \left(c_{2n}c_{3t} + c_{3t}c_{2n} + 2a_{2n}a_{3t}a_{3t}a_{2n}\right)$$

$$\times \frac{b_{2k}}{2}\frac{b_{3k}}{2} \sim (b_{1k} + b_{2k} + b_{3k})^2,$$

$$(f_k, f_r) = (f_k^2, f_r^2) + (f_k^3, f_r^3) = c_{2n}c_{2t}\frac{b_{2k}}{2}\frac{b_{2r}}{2} + c_{3n}c_{3t}\frac{b_{3k}}{2}\frac{b_{3r}}{2}$$

$$\sim b_{2k}b_{2r} + b_{3k}b_{3r}.$$

Finally, we have

$$(f_k, f_k) \sim (b_{1k} + b_{2k} + b_{3k})^2, \quad (f_k, f_r) \sim b_{2k}b_{2r} + b_{3k}b_{3r}, \quad k \neq r. \tag{5.4}$$

$$\text{Set} \quad \lambda_k = (b_{1k} + b_{2k} + b_{3k})^2 - (b_{2k}^2 + b_{3k}^2), \quad g_k = (b_{2k}, b_{3k}), \tag{5.5}$$

$$\text{then} \quad (f_k, f_k) \sim \lambda_k + (g_k, g_k), \quad (f_k, f_r) \sim (g_k, r_r). \tag{5.6}$$

For $A_{2m+1} = \left((f_k, f_r)\right)_{k,r=-m}^m$ and $b = -(b_{1k}/2)_{k=-m}^m \in \mathbb{R}^{2m+1}$, we have

$$A_{2m+1} = \sum_{k=-m}^m \lambda_k E_{kk} + \gamma(g_{-m}, \ldots, g_0, \ldots, g_m).$$

To complete the proof, it suffices to invoke Lemma 6.6 for $m = 2$. $\qquad \square$

5.2 Approximation of D_{rn} by A_{kn}

We formulate several useful lemmas for the approximation of the independent variables x_{kn} and operators D_{kn} using combinations of the generators A_{kn}. The generators A_{kn} have the following form:

$$A_{kn} = x_{1k}D_{1n} + x_{2k}D_{2n} + x_{3k}D_{3n}, \quad k, n \in \mathbb{Z}.$$

For $m = 3$, consider three rows as follows:

$$\begin{pmatrix} \cdots & a_{11} & a_{12} & \cdots & a_{1n} & \cdots \\ \cdots & a_{21} & a_{22} & \cdots & a_{2n} & \cdots \\ \cdots & a_{31} & a_{32} & \cdots & a_{2n} & \cdots \end{pmatrix} \quad \text{and set} \quad \lambda_k = \frac{1}{2b_{1k}} + \frac{1}{2b_{2k}} + \frac{1}{2b_{3k}}.$$

$$(5.7)$$

Denote by Y_1, Y_2 and Y_3 the three following vectors:

$$x_{rk}^a = \frac{a_{rk}}{\sqrt{\lambda_k}}, \quad k \in \mathbb{Z}, \quad Y_r = (x_{rk}^a)_{k \in \mathbb{Z}}. \tag{5.8}$$

The proofs of Lemmas 5.2 and 5.1 are based on Lemma 6.6 for $m = 2$.

Lemma 5.2. *For any $l \in \mathbb{Z}$, we have*

$$D_{rl}\mathbf{1} \in \langle A_{kl}\mathbf{1} \mid k \in \mathbb{Z} \rangle \quad \Leftrightarrow \quad \Delta(Y_r, Y_s, Y_t) = \infty,$$

where $\{r, s, t\}$ is a cyclic permutation of $\{1, 2, 3\}$.

Proof. Without loss of generality, we may assume that $r = 1$. We determine when the following holds:

$$D_{1n}\mathbf{1} \in \langle A_{kn}\mathbf{1} = (x_{1k}D_{1n} + x_{2k}D_{2n} + x_{3k}D_{3n})\mathbf{1} \mid k \in \mathbb{Z} \rangle.$$

Fix $m \in \mathbb{N}$; since $Mx_{1k} = a_{1k}$, we set $\sum_{k=-m}^{m} t_k a_{1k} = (t, b) = 1$, where $t = (t_k)_{k=-m}^{m}$ and $b = (a_{1k})_{k=-m}^{m}$. We have

$$\left\| \left[\sum_{k=-m}^{m} t_k(x_{1k}D_{1n} + x_{2k}D_{2n} + x_{3k}D_{3n}) - D_{1n} \right]\mathbf{1} \right\|^2$$

$$= \left\| \sum_{k=-m}^{m} t_k[(x_{1k} - a_{1k})D_{1n} + x_{2k}D_{2n} + x_{3k}D_{3n}]\mathbf{1} \right\|^2$$

$$= \sum_{-m \le k, r \le m} (f_k, f_r)t_k t_r =: (A_{2m+1}t, t),$$

where

$$A_{2m+1} = ((f_k, f_r))_{k,r=-m}^{m}, \text{ and } f_k = [(x_{1k} - a_{1k})D_{1n} + x_{2k}D_{2n} + x_{3k}D_{3n}]\mathbf{1}.$$

We get

$$(f_k, f_k) = \| \left[(x_{1k} - a_{1k})D_{1n} + x_{2k}D_{2n} + x_{3k}D_{3n} \right] \mathbf{1} \|^2$$

$$= \frac{1}{2b_{1k}} \frac{b_{1n}}{2} + \left(\frac{1}{2b_{2k}} + a_{2k}^2 \right) \frac{b_{2n}}{2} + \left(\frac{1}{2b_{3k}} + a_{3k}^2 \right) \frac{b_{3n}}{2}$$

$$\sim \frac{1}{2b_{1k}} + \frac{1}{2b_{2k}} + \frac{1}{2b_{3k}} + a_{2k}^2 + a_{3k}^2,$$

$$(f_k, f_r) = \left(\left[(x_{1k} - a_{1k})D_{1n} + x_{2k}D_{2n} + x_{3k}D_{3n} \right] \mathbf{1}, \left[(x_{1r} - a_{1r})D_{1n} \right. \right.$$

$$\left. \left. + x_{2r}D_{2n} + x_{3r}D_{3n} \right] \mathbf{1} \right)$$

$$= (x_{2k}\mathbf{1}, x_{2r}\mathbf{1})(D_{2n}\mathbf{1}, D_{2n}\mathbf{1})$$

$$+ (x_{3k}\mathbf{1}, x_{3r}\mathbf{1})(D_{3n}\mathbf{1}, D_{3n}\mathbf{1})$$

$$= a_{2k}a_{2r} \frac{b_{2n}}{2}$$

$$+ a_{3k}a_{3r} \frac{b_{3n}}{2} \sim a_{2k}a_{2r} + a_{3k}a_{3r}.$$

Finally, we have

$$(f_k, f_k) \sim \frac{1}{2b_{1k}} + \frac{1}{2b_{2k}} + \frac{1}{2b_{3k}} + a_{2k}^2 + a_{3k}^2,$$

$$(f_k, f_r) \sim a_{2k}a_{2r} + a_{3k}a_{3r}, \ k \neq r. \tag{5.9}$$

If we denote

$$\lambda_k = \frac{1}{2b_{1k}} + \frac{1}{2b_{2k}} + \frac{1}{2b_{3k}}, \quad g_k = (a_{2k}, a_{3k}), \tag{5.10}$$

then we have

$$(f_k, f_k) \sim \lambda_k + (g_k, g_k), \quad (f_k, f_r) \sim (g_k, g_r). \tag{5.11}$$

For $A_{2m+1} = \left((f_k, f_r) \right)_{k,r=-m}^{m}$ and $b = (a_{1k})_{k=-m}^{m} \in \mathbb{R}^{2m+1}$, we have

$$A_{2m+1} = \sum_{k=-m}^{m} \lambda_k E_{kk} + \gamma(g_{-m}, \ldots, g_0, \ldots, g_m).$$

To complete the proof, it suffices to apply Lemma 6.6 for $m = 2$. $\qquad \square$

5.3 Approximation of D_{kn} by $x_{rk}A_{kn}$

Set for $\{r, s, t\}$, a cyclic permutation of $\{1, 2, 3\}$:

$$\lambda_k^{(r)} = \left(\frac{1}{2b_{rk}} + a_{rk}^2\right)\left(C_k + \sum_{l=1, l\neq r}^{3} a_{lk}^2\right) - a_{rk}^2\left(\sum_{l=1, l\neq r}^{3} a_{lk}^2\right), \quad C_k = \sum_{l=1}^{3} \frac{1}{2b_{lk}},$$

$$(5.12)$$

$$Y_{rr} = \left(\frac{\frac{1}{2b_{rk}} + a_{rk}^2}{\sqrt{\lambda_k^{(r)}}}\right)_{k\in\mathbb{Z}}, Y_{rs} = \left(\frac{a_{rk}a_{sk}}{\sqrt{\lambda_k^{(r)}}}\right)_{k\in\mathbb{Z}}, \quad Y_{rt} = \left(\frac{a_{rk}a_{tk}}{\sqrt{\lambda_k^{(r)}}}\right)_{k\in\mathbb{Z}}.$$

$$(5.13)$$

Lemma 5.3. *For any $n \in \mathbb{Z}$ and $1 \leq r \leq 3$, we have*

$$D_{rn}\mathbf{1} \in \langle x_{1k}A_{kn}\mathbf{1} \mid k \in \mathbb{Z}\rangle \quad \Leftrightarrow \quad \Delta(Y_{rr}, Y_{rs}, Y_{rt}) = \infty,$$

where $\{r, s, t\}$ is a cyclic permutation of $\{1, 2, 3\}$.

Proof. We provide the proof for the case of $r = 1$. We determine when the following holds:

$$D_{1n}\mathbf{1} \in \langle x_{1k}A_{kn}\mathbf{1} = (x_{1k}^2 D_{1n} + x_{1k}x_{2k}D_{2n} + x_{1k}x_{3k}D_{3n})\mathbf{1} \mid k \in \mathbb{Z}\rangle.$$

Fix $m \in \mathbb{N}$; since $Mx_{1k}^2 = \frac{1}{2b_{1k}} + a_{1k}^2 =: c_{1k}$, we set $(t, b) = \sum_{k=-m}^{m} t_k c_{1k} = 1$, where $t = (t_k)_{k=-m}^{m}$ and $b = (c_{1k})_{k=-m}^{m}$. We have

$$\left\|\left[\sum_{k=-m}^{m} t_k(x_{1k}^2 D_{1n} + x_{1k}x_{2k}D_{2n} + x_{1k}x_{3k}D_{3n}) - D_{1n}\right]\mathbf{1}\right\|^2$$

$$= \left\|\sum_{k=-m}^{m} t_k\left[\left(x_{1k}^2 - c_{1k}\right)D_{1n} + x_{1k}x_{2k}D_{2n} + x_{1k}x_{3k}D_{3n}\right]\mathbf{1}\right\|^2$$

$$= \sum_{-m\leq k,r\leq m} (f_k, f_r)t_k t_r =: (A_{2m+1}t, t),$$

where

$$A_{2m+1} = ((f_k, f_r))_{k,r=-m}^{m}, \quad \text{and} \quad f_k = \left[\left(x_{1k}^2 - c_{1k}\right)D_{1n} + x_{1k}x_{2k}D_{2n} + x_{1k}x_{3k}D_{3n}\right]\mathbf{1}.$$

Since $M|\psi - M|\psi||^2 = M\psi^2 - |M\psi|^2$, we have $M\left|x_{1k}^2 - c_{1k}\right|^2 =$

$$Mx_{1k}^4 - c_{1k}^2 = \frac{3}{(2b_{1k})^2} + 6\frac{1}{2b_{1k}}a_{1k}^2 + a_{1k}^4 - c_{1k}^2 = \frac{1}{2b_{1k}}\left(\frac{2}{2b_{1k}} + 4a_{1k}^2\right).$$

Hence, we get

$$(f_k, f_k) = \left\|\left[(x_{1k}^2 - c_{1k})D_{1n} + x_{1k}x_{2k}D_{2n} + x_{1k}x_{3k}D_{3n}\right]\mathbf{1}\right\|^2$$

$$= \frac{1}{2b_{1k}}\left(\frac{2}{2b_{1k}} + 4a_{1k}^2\right)\frac{b_{1n}}{2} + c_{1k}c_{2k}\frac{b_{2n}}{2} + c_{1k}\frac{b_{3n}}{2}c_{3k}$$

$$\sim c_{1k}\left(C_k + a_{2k}^2 + a_{3k}^2\right),$$

$$(f_k, f_r) = \left(\left[\left(x_{1k}^2 - \left(\frac{1}{2b_{1k}} + a_{1k}^2\right)\right)D_{1n} + x_{1k}x_{2k}D_{2n} + x_{1k}x_{3k}D_{3n}\right]\mathbf{1},\right.$$

$$\left.\left[\left(x_{1r}^2 - \left(\frac{1}{2b_{1r}} + a_{1r}^2\right)\right)D_{1n} + x_{1r}x_{2r}D_{2n} + x_{1r}x_{3r}D_{3n}\right]\mathbf{1}\right)$$

$$= (x_{1k}\mathbf{1}, x_{1r}\mathbf{1})(x_{2k}\mathbf{1}, x_{2r}\mathbf{1})\|D_{2n}\mathbf{1}\|^2 + (x_{1k}\mathbf{1}, x_{1r}\mathbf{1})(x_{3k}\mathbf{1}, x_{3r}\mathbf{1})\|D_{3n}\mathbf{1}\|^2$$

$$= a_{1k}a_{1r}a_{2k}a_{2r}\frac{b_{2n}}{2} + a_{1k}a_{1r}a_{3k}a_{3r}\frac{b_{3n}}{2}$$

$$\simeq a_{1k}a_{1r}(a_{2k}a_{2r} + a_{3k}a_{3r}).$$

Finally, we have

$$(f_k, f_k) \sim \left(\frac{1}{2b_{1k}} + a_{1k}^2\right)\left(\frac{1}{2b_{1k}} + \frac{1}{2b_{2k}} + \frac{1}{2b_{3k}} + a_{2k}^2 + a_{3k}^2\right),$$

$$(f_k, f_r) \sim a_{1k}a_{1r}(a_{2k}a_{2r} + a_{3k}a_{3r}), \ \ k \neq r. \tag{5.14}$$

Set

$$\lambda_k^{(1)} = \left(\frac{1}{2b_{1k}} + a_{1k}^2\right)\left(\frac{1}{2b_{1k}} + \frac{1}{2b_{2k}} + \frac{1}{2b_{3k}} + a_{2k}^2 + a_{3k}^2\right)$$

$$- a_{1k}^2(a_{2k}^2 + a_{3k}^2), \tag{5.15}$$

$$g_k = a_{1k}(a_{2k}, a_{3k}), \quad \text{then}$$

$$(f_k, f_k) = \lambda_k^{(1)} + (g_k, g_k), \quad (f_k, f_r) \sim (g_k, g_r), \ \ k \neq r.$$

For $A_{2m+1} = ((f_k, f_r))_{k,r=-m}^m$ and $b = (a_{1k})_{k=-m}^m \in \mathbb{R}^{2m+1}$, we have

$$A_{2m+1} = \sum_{k=-m}^m \lambda_k E_{kk} + \gamma(g_{-m}, \ldots, g_0, \ldots, g_m).$$

To complete the proof, it suffices to apply Lemma 6.6 for $m = 2$. $\qquad\square$

5.4   Approximation of D_{rn} by $\exp\big(is_k(x_{rk} - a_{rk})\big)A_{kn}$

Lemma 5.4. *We have*

$$D_{3k}\mathbf{1} \in \big\langle \sin\big(s_k(x_{3k} - a_{3k})\big)A_{kn}\mathbf{1} \mid k \in \mathbb{Z}\big\rangle \;\Leftrightarrow\; \Sigma_3(D, s) = \infty,$$
$$(5.16)$$

$$D_{3k}\mathbf{1} \in \big\langle \cos\big(s_k(x_{3k} - a_{3k})\big)A_{kn}\mathbf{1} \mid k \in \mathbb{Z}\big\rangle \;\Leftrightarrow\; \Sigma_3^\vee(D, s) = \infty,$$
$$(5.17)$$

where $\Sigma_3(D, s) = \sum_{k \in \mathbb{Z}} \dfrac{|M\eta_{3k}(s_k)|^2}{\|g_k(s_k)\|^2}, \;\; \Sigma_3^\vee(D, s) = \sum_{k \in \mathbb{Z}} \dfrac{|M\eta_{3k}^\vee(s_k)|^2}{\|g_k^\vee(s_k)\|^2},$

$$(5.18)$$

moreover, $\quad \Sigma_3(D, s^{(3)}) \sim \Sigma_3(D) := \sum_k \dfrac{\frac{1}{2b_{3k}}}{C_k + a_{1k}^2 + a_{2k}^2 + a_{3k}^2},$

$$(5.19)$$

and $\quad \Sigma_3^\vee(D, s^{(3)}) \sim \Sigma_3^\vee(D) := \sum_k \dfrac{a_{3k}^2}{C_k + a_{1k}^2 + a_{2k}^2 + a_{3k}^2}, \quad (5.20)$

where $s^{(3)} = (s_{3k})_k,$ *with* $\frac{s_{3k}^2}{b_{3k}} \equiv 1, \; k \in \mathbb{Z}$ *and*

$$\eta_{3k}(s_k), \; \eta_{3k}^\vee(s_k), \; g_k(s_k), \; g_k^\vee(s_k)$$

being defined by (5.26)–(5.28).

Proof.   We shall try to obtain separately the real and imaginary parts of $M\xi_{3k}(s)$, where $\xi_{3k}(s_k) = ix_{3k}\exp\big(is_k(x_{3k} - a_{3k})\big)$. Setting

$$F_b(s) = \int_{\mathbb{R}} \exp\big(is(x - a)\big)d\mu_{(b,a)}(x),$$

we obtain

$$F_b(s) = \int_{\mathbb{R}} \exp\big(isx\big)d\mu_{(b,0)}(x) = \exp\left(-\frac{s^2}{4b}\right), \qquad (5.21)$$

where $d\mu_{(b,a)}(x)$ and $d\mu_{(b,0)}(x)$ are defined by

$$d\mu_{(b,a)}(x) = \sqrt{\frac{b}{\pi}}\, e^{-b(x-a)^2}\, dx \quad \text{and} \quad d\mu_{(b,0)}(x) = \sqrt{\frac{b}{\pi}}\, e^{-bx^2}\, dx. \quad (5.22)$$

Therefore,

$$H_{a,b}(s) = \int_{\mathbb{R}} ixe^{is(x-a)}\, d\mu_{(b,a)}(x) = \int_{\mathbb{R}} i(x+a)e^{isx}\, d\mu_{(b,0)}(x) \quad (5.23)$$

$$= \frac{dF_b(s)}{ds} + iaF_b(s) = \left(-\frac{s}{2b} + ia\right)\exp\left(-\frac{s^2}{4b}\right). \quad (5.24)$$

Recall the Euler formulas:

$$e^{it} = \cos t + i\sin t, \quad e^{-it} = \cos t - i\sin t,$$

$$\cos t = \frac{e^{it} + e^{-it}}{2}, \quad \sin t = \frac{e^{it} - e^{-it}}{2i}. \quad (5.25)$$

More precisely, we denote for $1 \le r \le 3$,

$$\eta_{rk}(s) = x_{rk}\cos\big(s_k(x_{3k} - a_{3k})\big), \quad \overset{\vee}{\eta}_{rk}(s) = x_{rk}\cos\big(s_k(x_{3k} - a_{3k})\big). \quad (5.26)$$

We determine when the inclusion holds:

$$D_{3k}\mathbf{1} \in \big\langle \sin\big(s_k(x_{3k} - a_{3k})\big)A_{kn}\mathbf{1}$$

$$= \Big(x_{1k}\sin\big(s_k(x_{3k} - a_{3k})\big)D_{1n} + x_{2k}\sin\big(s_k(x_{3k} - a_{3k})\big)D_{2n}$$

$$+ x_{3k}\sin\big(s_k(x_{3k} - a_{3k})\big)D_{3n}\Big)\mathbf{1} \mid k \in \mathbb{Z}\big\rangle,$$

$$D_{3k}\mathbf{1} \in \big\langle \cos\big(s_k(x_{3k} - a_{3k})\big)A_{kn}\mathbf{1}$$

$$= \Big(x_{1k}\cos\big(s_k(x_{3k} - a_{3k})\big)D_{1n} + x_{2k}\cos\big(s_k(x_{3k} - a_{3k})\big)D_{2n}$$

$$+ x_{3k}\cos\big(s_k(x_{3k} - a_{3k})\big)D_{3n}\Big)\mathbf{1} \mid k \in \mathbb{Z}\big\rangle.$$

Set

$$g_k(s_k) = \Big(\eta_{1k}(s_k)D_{1n} + \eta_{2k}(s_k)D_{2n} + \big[\eta_{3k}(s_k) - M\eta_{3k}(s_k)\big]D_{3n}\Big)\mathbf{1}, \quad (5.27)$$

$$\overset{\vee}{g}_k(s_k) = \Big(\overset{\vee}{\eta}_{1k}(s_k)D_{1n} + \overset{\vee}{\eta}_{2k}(s_k)D_{2n} + \big[\overset{\vee}{\eta}_{3k}(s_k) - M\overset{\vee}{\eta}_{3k}(s_k)\big]D_{3n}\Big)\mathbf{1}. \quad (5.28)$$

We show that

$$M\eta_{3k}(s) = -\frac{1}{2}\Big(H_{a,b}(s) + \overline{H_{a,b}(s)}\Big) = \frac{s}{2b_{3k}}\exp\Big(-\frac{s^2}{4b_{3k}}\Big), \quad (5.29)$$

$$M\eta_{3k}^{\vee}(s) = \frac{1}{2i}\Big(H_{a,b}(s) - \overline{H_{a,b}(s)}\Big) = a_{3k}\exp\Big(-\frac{s^2}{4b_{3k}}\Big). \quad (5.30)$$

Using the function $F_b(s)$, defined by (5.21), we get

$$M\eta(s) = \int_{\mathbb{R}} x\sin\big(s(x-a)\big)d\mu_{(b,a)}(x) = \int_{\mathbb{R}}(x+a)\sin(sx)d\mu_{(b,0)}(x)$$

$$= \int_{\mathbb{R}}(x+a)\frac{e^{isx} - e^{-isx}}{2i}d\mu_{(b,0)}(x)$$

$$= -\frac{1}{2}\int_{\mathbb{R}} i(x+a)\big(e^{isx} - e^{-isx}\big)d\mu_{(b,0)}(x)$$

$$= -\frac{1}{2}\Big(H_{a,b}(s) + \overline{H_{a,b}(s)}\Big) = \frac{s}{2b}\exp\Big(-\frac{s^2}{4b}\Big),$$

which implies (5.30). Similarly, we get

$$M\eta^{\vee}(s) = \int_{\mathbb{R}} x\cos\big(s(x-a)\big)d\mu_{(b,a)}(x)$$

$$= \int_{\mathbb{R}}(x+a)\cos(sx)d\mu_{(b,0)}(x)$$

$$= \frac{1}{2i}\int_{\mathbb{R}} i(x+a)\Big(e^{isx} + e^{-isx}\Big)d\mu_{(b,0)}(x)$$

$$= \frac{1}{2i}\Big(H_{a,b}(s) - \overline{H_{a,b}(s)}\Big) = ae^{-\frac{s^2}{4b}},$$

which implies (5.29). Fix $m \in \mathbb{N}$, and we set $\sum_{k=-m}^{m} t_k M\eta_{3k}(s_k) = (t,b) = 1$, where $t = (t_k)_{k=-m}^{m}$ and $b = (M\eta_{3k}(s_k))_{k=-m}^{m}$. We have

$$\left\|\left[\sum_{k=-m}^{m} t_k \sin\big(s_k(x_{3k} - a_{3k})\big)A_{kn} - D_{3n}\right]\mathbf{1}\right\|^2$$

$$= \left\|\sum_{k=-m}^{m} t_k\Big(\eta_{1k}(s_k)D_{1n} + \eta_{2k}(s_k)D_{2n}\right.$$

$$\left. + \big[\eta_{3k}(s_k) - M\eta_{3k}(s_k)\big]D_{3n}\Big)\mathbf{1}\right\|^2$$

$$= \sum_{k=-m}^{m} t_k^2 \big\| g_k(s_k) \big\|^2, \ \ \text{since} \ \Big(D_{rn}\mathbf{1}, D_{ln}\mathbf{1} \Big) = 0, \ \ 1 \le r < l \le 3,$$

$$(5.31)$$

where $g_k(s_k)$ are defined by (5.27). In order to calculate $\|g_k(s_k)\|^2$, note that

$$\|g_k(s_k)\|^2 = (g_k(s_k), g_k(s_k)) = \Big(\big(\eta_{1k}(s_k)D_{1n} + \eta_{2k}(s_k)D_{2n}$$

$$+ \big[\eta_{3k}(s_k) - M\eta_{3k}(s_k) \big] D_{3n} \big) \mathbf{1},$$

$$\big(\eta_{1k}(s_k)D_{1n} + \eta_{2k}(s_k)D_{2n} + \big[\eta_{3k}(s_k) - M\eta_{3k}(s_k) \big] D_{3n} \big) \mathbf{1} \Big)$$

$$= \|x_{1k}\mathbf{1}\|^2 \| \sin \big(s_k(x_{3k} - a_{3k}) \big) \mathbf{1} \|^2 \|D_{1k}\mathbf{1}\|^2$$

$$+ \|x_{2k}\mathbf{1}\|^2 \| \sin \big(s_k(x_{3k} - a_{3k}) \big) \mathbf{1} \|^2 \|D_{2k}\mathbf{1}\|^2$$

$$+ \Big(M|\eta_{kn}(s_k)|^2 - |M\eta_{kn}(s_k)|^2 \Big) \|D_{3k}\mathbf{1}\|^2$$

$$= \Big(\frac{1}{2b_{1k}} + a_{1k}^2 \Big) I_3 \frac{b_{1n}}{2} + \Big(\frac{1}{2b_{2k}} + a_{2k}^2 \Big) I_3 \frac{b_{2n}}{2}$$

$$+ \Big(M|\eta_{kn}(s_k)|^2 - |M\eta_{kn}(s_k)|^2 \Big) \frac{b_{3n}}{2}. \qquad (5.32)$$

We need to calculate $I_3 = \| \sin \big(s_k(x_{3k} - a_{3k}) \big) \mathbf{1} \|^2$, $M|\eta_{kn}(s_k)|^2$ and $|M\eta_{kn}(s_k)|^2$. If we set $a := a_{3k}$, $b := b_{3k}$, we get

$$I_3 = \| \sin \big(s_k(x_{3k} - a_{3k}) \big) \mathbf{1} \|^2$$

$$= \int_{\mathbb{R}} \frac{e^{isx} - e^{-isx}}{2i} \frac{e^{-isx} - e^{isx}}{-2i} d\mu_{(b,0)}(x)$$

$$= \frac{1}{2} \int_{\mathbb{R}} \Big(1 - \frac{e^{2isx} + e^{-2isx}}{2} \Big) d\mu_{(b,0)}(x) \overset{(5.21)}{=} \frac{1 - e^{-\frac{s^2}{b}}}{2},$$

$$(5.33)$$

$$|M\eta_{kn}(s_k)|^2 = \frac{s_k^2}{4b_{3k}^2}\exp\left(-\frac{s_k^2}{2b_{3k}}\right), \tag{5.34}$$

$$M|\eta_{kn}(s_k)|^2 = \int_{\mathbb{R}}(x^2 + 2xa + a^2)\frac{e^{isx}-e^{-isx}}{2i}\frac{e^{-isx}-e^{isx}}{-2i}d\mu_{(b,0)}(x)$$

$$= \frac{1}{2}\int_{\mathbb{R}}(x^2 + 2xa + a^2)\left(1 - \frac{e^{2isx}+e^{-2isx}}{2}\right)d\mu_{(b,0)}(x)$$

$$= \frac{1}{2}\left[\int_{\mathbb{R}}(x^2 + a^2)d\mu_{(b,0)}(x)\right.$$

$$\left. - \int_{\mathbb{R}}(x^2 + a^2)\frac{e^{2isx}+e^{-2isx}}{2}d\mu_{(b,0)}(x)\right]$$

$$= \frac{1}{2}\left[\frac{1}{2b} + a^2 - \frac{d^2 F_b(2s)}{ds^2} - a^2 F_b(2s)\right]$$

$$\overset{(5.21)}{=} \frac{1}{2}\left[\frac{1}{2b} + a^2 - \frac{1}{(2i)^2}\left[\left(\frac{2s}{b}\right)^2 - \frac{2}{b}\right]e^{-\frac{s^2}{b}} - a^2 e^{-\frac{s^2}{b}}\right]$$

$$= \frac{1}{2}\left[\left(\frac{1}{2b} + a^2\right)(1 - e^{-\frac{s^2}{b}}) + \frac{s^2}{b^2}e^{-\frac{s^2}{b}}\right]. \tag{5.35}$$

Finally, we get

$$M|\eta_{kn}(s_k)|^2 - |M\eta_{kn}(s_k)|^2$$

$$= \frac{1}{2}\left[\left(\frac{1}{2b} + a^2\right)(1 - e^{-\frac{s^2}{b}}) + \frac{s^2}{b^2}e^{-\frac{s^2}{b}}\right] - \frac{s^2}{4b^2}e^{-\frac{s^2}{2b}}. \tag{5.36}$$

By (5.32), (5.33), (5.36), (5.37) and (6.2), we prove (5.16), where

$$\Sigma_3(D,s) = \sum_{k\in\mathbb{Z}}\frac{|M\eta_{kn}(s_k)|^2}{\|g_k(s_k)\|^2}$$

$$= \sum_{k\in\mathbb{Z}}\frac{\frac{s_k^2}{4b_{3k}^2}e^{-\frac{s_k^2}{2b_{3k}}}}{\left(\frac{1}{2b_{1k}} + a_{1k}^2\right)I_3\frac{b_{1n}}{2} + \left(\frac{1}{2b_{2k}} + a_{2k}^2\right)I_3\frac{b_{2n}}{2} + \left(M|\eta_{kn}(s_k)|^2 - |M\eta_{kn}(s_k)|^2\right)\frac{b_{3n}}{2}}$$

$$\sim \sum_{k\in\mathbb{Z}} \frac{\frac{s_k^2}{4b_{3k}^2}e^{-\frac{s_k^2}{2b_{3k}}}}{\frac{1-e^{-\frac{s_k^2}{b_{3k}}}}{2}\left(c_{1k}+c_{2k}\right)+\frac{1}{2}\left[c_{3k}(1-e^{-\frac{s_k^2}{b_{3k}}})+\frac{s_k^2}{b_{3k}^2}e^{-\frac{s_k^2}{b_{3k}}}\right]-\frac{s_k^2}{4b_{3k}^2}e^{-\frac{s_k^2}{2b_{3k}}}}$$

$$=\sum_{k\in\mathbb{Z}} \frac{\frac{x_k^2}{4b_{3k}}e^{-\frac{x_k^2}{2}}}{\frac{1-e^{-x_k^2}}{2}\left(c_{1k}+c_{2k}\right)+\frac{1}{2}\left[c_{3k}(1-e^{-x_k^2})+\frac{x_k^2}{b_{3k}}e^{-x_k^2}\right]-\frac{x_k^2}{4b_{3k}}e^{-\frac{x_k^2}{2}}} =: \Sigma_3(D,x),$$

where $x_k^2=\frac{s_k^2}{b_{3k}}$ and $c_{rk}=\frac{1}{2b_{rk}}+a_{rk}^2$. For $x^{(3)}=(x_k)_k$, with $x_k\equiv 1$, we get

$$\Sigma_3(D,x^{(3)})$$

$$=\sum_{k\in\mathbb{Z}} \frac{\frac{1}{4b_{3k}}e^{-\frac{1}{2}}}{\frac{1-e^{-1}}{2}\left(c_{1k}+c_{2k}\right)+\frac{1}{2}\left[c_{3k}(1-e^{-1})+\frac{1}{b_{3k}}e^{-1}\right]-\frac{1}{4b_{3k}}e^{-\frac{1}{2}}}$$

$$=\sum_{k\in\mathbb{Z}} \frac{\frac{1}{4b_{3k}}e^{-\frac{1}{2}}}{\frac{1-e^{-1}}{2}\left(c_{1k}+c_{2k}+c_{3k}\right)+\frac{1}{2b_{3k}}\left(e^{-1}-\frac{1}{2}e^{-\frac{1}{2}}\right)}$$

$$\overset{(3.6)}{\sim}\sum_{k\in\mathbb{Z}}\frac{\frac{1}{2b_{3k}}}{c_{1k}+c_{2k}+c_{3k}}=\sum_{k\in\mathbb{Z}}\frac{\frac{1}{2b_{3k}}}{C_k+a_{1k}^2+a_{2k}^2+a_{3k}^2}=\Sigma_3(D).$$

$$(5.37)$$

So, we have proved (5.19) for $x=(x_k)_k$, with $x_k\equiv 1$. To approximate D_{3n} in terms of functions involving cosine, fix $m\in\mathbb{N}$, and set $\sum_{k=-m}^{m}t_k M\eta_{3k}^{\vee}(s_k)=(t,b)=1$, where $t=(t_k)_{k=-m}^{m}$ and $b=(M\eta_{3k}^{\vee}(s_k))_{k=-m}^{m}$. We have

$$\left\|\left[\sum_{k=-m}^{m}t_k\cos\left(s_k(x_{3k}-a_{3k})\right)A_{kn}-D_{3n}\right]\mathbf{1}\right\|^2$$

$$=\left\|\sum_{k=-m}^{m}t_k\left(\eta_{1k}^{\vee}(s_k)D_{1n}+\eta_{2k}^{\vee}(s_k)D_{2n}\right.\right.$$

$$\left.\left.+\left[\eta_{3k}^{\vee}(s_k)-M\eta_{3k}^{\vee}(s_k)\right]D_{3n}\right)\mathbf{1}\right\|^2$$

$$= \sum_{k=-m}^{m} t_k^2 \|g_k^{\vee}(s_k)\|^2, \ \text{ since } \ \left(D_{rn}\mathbf{1}, D_{ln}\mathbf{1}\right) = 0, \ \ 1 \leq r < l \leq 3,$$

$$(5.38)$$

where $g_k^{\vee}(s_k)$ are defined by (5.27). To calculate $\|g_k^{\vee}(s_k)\|^2$, we have

$$\|g_k^{\vee}(s_k)\|^2 = (g_k^{\vee}(s_k), g_k^{\vee}(s_k))$$
$$= \Big(\big(\eta_{1k}^{\vee}(s_k) D_{1n} + \eta_{2k}^{\vee}(s_k) D_{2n} + \big[\eta_{3k}^{\vee}(s_k) - M\eta_{3k}^{\vee}(s_k) \big] D_{3n} \big) \mathbf{1},$$
$$\big(\eta_{1k}^{\vee}(s_k) D_{1n} + \eta_{2k}^{\vee}(s_k) D_{2n} + \big[\eta_{3k}^{\vee}(s_k) - M\eta_{3k}^{\vee}(s_k) \big] D_{3n} \big) \mathbf{1} \Big)$$
$$= \|x_{1k}\mathbf{1}\|^2 \|\cos\big(s_k(x_{3k} - a_{3k})\big)\mathbf{1}\|^2 \|D_{1k}\mathbf{1}\|^2$$
$$+ \|x_{2k}\mathbf{1}\|^2 \|\cos\big(s_k(x_{3k} - a_{3k})\big)\mathbf{1}\|^2 \|D_{2k}\mathbf{1}\|^2$$
$$+ \Big(M|\eta_{kn}^{\vee}(s_k)|^2 - |M\eta_{kn}^{\vee}(s_k)|^2 \Big) \|D_{3k}\mathbf{1}\|^2$$
$$= \Big(\frac{1}{2b_{1k}} + a_{1k}^2 \Big) I_3^{\vee} \frac{b_{1n}}{2} + \Big(\frac{1}{2b_{2k}} + a_{2k}^2 \Big) I_3^{\vee} \frac{b_{2n}}{2}$$
$$+ \Big(M|\eta_{kn}^{\vee}(s_k)|^2 - |M\eta_{kn}^{\vee}(s_k)|^2 \Big) \frac{b_{3n}}{2}. \qquad (5.39)$$

We need to calculate $I_3^{\vee} = \|\cos\big(is_k(x_{3k} - a_{3k})\big)\mathbf{1}\|^2$, $M|\eta_{kn}^{\vee}(s_k)|^2$ and $|M\eta_{kn}^{\vee}(s_k)|^2$. If we set $a := a_{3k}, \ b := b_{3k}$, we get

$$\begin{aligned}
I_3^{\vee} &= \|\cos\big(s_k(x_{3k} - a_{3k})\big)\mathbf{1}\|^2 \\
&= \int_{\mathbb{R}} \frac{e^{isx} + e^{-isx}}{2} \frac{e^{-isx} + e^{isx}}{2} d\mu_{(b,0)}(x) \\
&= \frac{1}{2} \int_{\mathbb{R}} \Big(1 + \frac{e^{2isx} + e^{-2isx}}{2} \Big) d\mu_{(b,0)}(x) \\
&\overset{(5.21)}{=} \frac{1 + e^{-\frac{s^2}{b}}}{2},
\end{aligned} \qquad (5.40)$$

$$|M\eta^{\vee}_{kn}(s_k)|^2 \;=\; a^2_{3k}\exp\Big(-\frac{s^2_k}{2b_{3k}}\Big), \tag{5.41}$$

$$M|\eta^{\vee}_{kn}(s_k)|^2 \;=\; \int_{\mathbb{R}}(x^2+2xa+a^2)\frac{e^{isx}+e^{-isx}}{2}\frac{e^{-isx}+e^{isx}}{2}d\mu_{(b,0)}(x)$$

$$=\; \frac{1}{2}\int_{\mathbb{R}}(x^2+2xa+a^2)\Big(1+\frac{e^{2isx}+e^{-2isx}}{2}\Big)d\mu_{(b,0)}(x)$$

$$=\; \frac{1}{2}\Big[\int_{\mathbb{R}}(x^2+a^2)d\mu_{(b,0)}(x)+\int_{\mathbb{R}}(x^2+a^2)$$

$$\times\;\frac{e^{2isx}+e^{-2isx}}{2}d\mu_{(b,0)}(x)\Big]$$

$$=\; \frac{1}{2}\Big[\frac{1}{2b}+a^2+\frac{d^2 F_b(2s)}{ds^2}+a^2 F_b(2s)\Big]$$

$$\overset{(5.21)}{=}\; \frac{1}{2}\Big[\frac{1}{2b}+a^2+\frac{1}{(2i)^2}\Big[\Big(\frac{2s}{b}\Big)^2-\frac{2}{b}\Big]e^{-\frac{s^2}{b}}+a^2 e^{-\frac{s^2}{b}}\Big]$$

$$=\; \frac{1}{2}\Big[\Big(\frac{1}{2b}+a^2\Big)(1+e^{-\frac{s^2}{b}})-\frac{s^2}{b^2}e^{-\frac{s^2}{b}}\Big]. \tag{5.42}$$

Finally, we get

$$M|\eta^{\vee}_{kn}(s_k)|^2-|M\eta^{\vee}_{kn}(s_k)|^2=\frac{1}{2}\Big[\Big(\frac{1}{2b}+a^2\Big)(1+e^{-\frac{s^2}{b}})-\frac{s^2}{b^2}e^{-\frac{s^2}{b}}\Big]-e^{-\frac{s^2}{2b}}. \tag{5.43}$$

By (5.38), (5.39), (5.40), (5.43) and (6.2), we prove (5.17), where

$$\Sigma^{\vee}_3(D,s)=\sum_{k\in\mathbb{Z}}\frac{|M\eta^{\vee}_{kn}(s_k)|^2}{\|g^{\vee}_k(s_k)\|^2}$$

$$=\sum_{k\in\mathbb{Z}}\frac{a^2_{3k}\exp(-s^2_k/2b_{3k})}{\Big(\frac{1}{2b_{1k}}+a^2_{1k}\Big)I^{\vee}_3\frac{b_{1n}}{2}+\Big(\frac{1}{2b_{2k}}+a^2_{2k}\Big)I^{\vee}_3\frac{b_{2n}}{2}+\Big(M|\eta^{\vee}_{kn}(s_k)|^2-|M\eta^{\vee}_{kn}(s_k)|^2\Big)\frac{b_{3n}}{2}}$$

$$\sim\sum_{k\in\mathbb{Z}}\frac{a^2_{3k}\exp(-\frac{s^2_k}{2b_{3k}})}{\frac{1+\exp(-\frac{s^2_k}{b_{3k}})}{2}\Big(c_{1k}+c_{2k}\Big)+\frac{1}{2}\Big[c_{3k}(1+\exp(-\frac{s^2_k}{b_{3k}}))-\frac{s^2_k}{b^2_{3k}}\exp(-\frac{s^2_k}{b_{3k}})\Big]-a^2_{3k}\exp(-\frac{s^2_k}{2b_{3k}})}$$

$$=\sum_{k\in\mathbb{Z}}\frac{a^2_{3k}\exp(-x^2_k/2)}{\frac{1-e^{-x^2_k}}{2}\Big(c_{1k}+c_{2k}\Big)+\frac{1}{2}\Big[c_{3k}(1+e^{-x^2_k})-\frac{x^2_k}{b_{3k}}e^{-x^2_k}\Big]-a^2_{3k}e^{-\frac{x^2_k}{2}}}=\Sigma_3(D,x), \tag{5.44}$$

where $x_k^2 = \frac{s_k^2}{b_{3k}}$ and $c_{rk} = \frac{1}{2b_{rk}} + a_{rk}^2$. For $x^{(3)} = (x_k)_k$, with $x_k \equiv 1$, we get

$$\Sigma_3^{\vee}(D, x^{(3)})$$

$$= \sum_{k \in \mathbb{Z}} \frac{a_{3k}^2 e^{-\frac{1}{2}}}{\frac{1+e^{-1}}{2}\left(c_{1k} + c_{2k}\right) + \frac{1}{2}\left[c_{3k}(1 + e^{-1}) - \frac{1}{b_{3k}}e^{-1}\right] - a_{3k}^2 e^{-\frac{1}{2}}}$$

$$= \sum_{k \in \mathbb{Z}} \frac{a_{3k}^2 e^{-\frac{1}{2}}}{\frac{1+e^{-1}}{2}\left(c_{1k} + c_{2k} + c_{3k}\right) - \left(\frac{1}{2b_{3k}}e^{-1} + a_{3k}^2 e^{-\frac{1}{2}}\right)}$$

$$\overset{(3.6)}{\sim} \sum_{k \in \mathbb{Z}} \frac{a_{3k}^2}{c_{1k} + c_{2k} + c_{3k}} = \sum_{k \in \mathbb{Z}} \frac{a_{3k}^2}{C_k + a_{1k}^2 + a_{2k}^2 + a_{3k}^2} = \Sigma_3^{\vee}(D).$$

$$(5.45)$$

So, we have proved (5.20) for $x = (x_k)_k$, with $x_k \equiv 1$. $\qquad\square$

Chapter 6

Some Useful Facts

6.1 Some Estimates

We obtain the well-known estimates (see, e.g., Beckenbach and Bellman Chap. I, §52):

$$\min_{x\in\mathbb{R}^n}\left(\sum_{k=1}^{n}a_kx_k^2\;\middle|\;\sum_{k=1}^{n}x_k=1\right)=\left(\sum_{k=1}^{n}\frac{1}{a_k}\right)^{-1},\quad a_k>0,\quad x_k\in\mathbb{R}.\tag{6.1}$$

We also use the same estimate in a slightly different form:

$$\min_{x\in\mathbb{R}^n}\left(\sum_{k=1}^{n}a_kx_k^2\;\middle|\;\sum_{k=1}^{n}x_kb_k=1\right)=\left(\sum_{k=1}^{n}\frac{b_k^2}{a_k}\right)^{-1}.\tag{6.2}$$

The minimum is obtained for $x_k=\frac{b_k}{a_k}\left(\sum_{k=1}^{n}\frac{b_k^2}{a_k}\right)^{-1}$.

Lemma 6.1 (Kosyak, 2004, 2018). *For a strictly positive operator A, satisfying $(Af,f)>0$, $f\neq 0$, acting on $\mathbb{R}^n$ and a vector $b\in\mathbb{R}^n\backslash\{0\}$, we have*

$$\min_{x\in\mathbb{R}^n}\{(Ax,x)\mid(x,b)=1\}=\frac{1}{(A^{-1}b,b)}.\tag{6.3}$$

The minimum is obtained for $x=\frac{1}{(A^{-1}b,b)}A^{-1}b$.

Lemma 6.1 is a direct generalisation of (6.2). Denote by $D(B)$ the domain of the definition of an operator B acting in some Hilbert space.

107

Lemma 6.2 (Kosyak, 2025a). *For a strictly positive operator A on an infinite-dimensional Hilbert space H and a vector $b \in H\backslash\{0\}$ such that $b \in D(A^{-1})$, we have*

$$\min_{x \in H}\{(Ax, x) \mid (x, b) = 1\} = \frac{1}{(A^{-1}b, b)}. \tag{6.4}$$

The minimum is reached for $x_0 = \frac{1}{(A^{-1}b,b)}A^{-1}b$.

Proof.   Consider a new scalar product in H, defined as follows:

$$(f, g)_A := (Af, g), \quad f, g \in H. \tag{6.5}$$

Since

$$(Ax, x) = (x, x)_A = \|x\|_A^2 \quad \text{and} \quad 1 = (b, x) = (A^{-1}b, x)_A,$$

the minimum $\|x\|_A^2$ will be achieved on the vector $x_0 = sA^{-1}b$, generating the hyperplane $1 = (A^{-1}b, x)_A$ and belonging to this hyperplane. We get

$$1 = (b, sA^{-1}b), \quad \text{therefore} \quad s = \frac{1}{(A^{-1}b, b)}, \quad x_0 = \frac{1}{(A^{-1}b, b)}A^{-1}b.$$

Finally, we get $(Ax_0, x_0) = \frac{1}{(A^{-1}b,b)}$. $\square$

Example 6.1. For a positive definite operator $A = \mathrm{diag}(\lambda_k)_{k=1}^{\infty}$ in $l_2(\mathbb{N})$, where $\lambda_k = \frac{1}{k}$ and $b = (b_k)_{k \in \mathbb{N}} \in l_2(\mathbb{N})$ with $b_k = \frac{1}{k}$, we have $b \notin D(A^{-1})$ since $(A^{-1}b)_k \equiv 1$ for all $k \in \mathbb{N}$. Hence, $A^{-1}b \notin l_2(\mathbb{N})$. In this case, $\frac{1}{(A^{-1}b,b)} = 0$. Indeed, for the corresponding projections A_n, b_n on $\mathbb{R}^n$, we have

$$(A_n^{-1}b_n, b_n) = \sum_{k=1}^{n} \frac{1}{k} \to \infty.$$

6.2   How Far Is a Vector from a Hyperplane?

We start with a classical result; see, e.g., Gantmacher (1958). Consider the hyperplane V_n generated by n arbitrary vectors $f_1, \ldots, f_n$ in some Hilbert space H.

Lemma 6.3 (Akhiezer and Glazman, 1993; Gantmacher, 1958).
The square of the distance $d(f_0, V_n)$ of a vector f_0 from the hyperplane V_n is given by the ratio of two Gram determinants (see Definition 7.3):

$$d^2(f_0, V_n) = \frac{\Gamma(f_0, f_1, f_2, \ldots, f_n)}{\Gamma(f_1, f_2, \ldots, f_n)}. \tag{6.6}$$

Proof. We follow closely the book by Akhiezer and Glazman (1993). Set $f = \sum_{k=1}^{n} t_k f_k \in V_n$ and $h = f - f_0$. Since h should be orthogonal to V_n, we conclude that $f_r \perp h$, i.e., $(f_r, h) = 0$ for all r, or

$$\sum_{k=1}^{n} t_k(f_r, f_k) = (f_r, f_0), \quad 1 \leq r \leq n. \tag{6.7}$$

Set $A = \gamma(f_1, f_2, \ldots, f_n)$ and $b = (f_k, f_0)_{k=1}^{n} \in \mathbb{R}^n$. By definition, we have

$$d^2 = \min_{f \in V_n} \|f - f_0\|^2 = (At, t) - 2(t, b) + (f_0, f_0). \tag{6.8}$$

Since $d^2 = (h, h) = (f_0, h)$, we conclude that $d^2 = \sum_{k=1}^{n} t_k(f_0, f_k) - (f_0, f_0)$, or

$$\sum_{k=1}^{n} t_k(f_0, f_k) = (f_0, f_0) - d^2. \tag{6.9}$$

So, we have the following system of equations:

$$\begin{cases} t_1(f_1, f_1) + t_2(f_1, f_2) + \cdots + t_n(f_1, f_n) = (f_1, f_0) \\ t_1(f_2, f_1) + t_2(f_2, f_2) + \cdots + t_n(f_2, f_n) = (f_2, f_0) \\ \qquad\qquad\qquad \vdots \\ t_1(f_n, f_1) + t_2(f_n, f_2) + \cdots + t_n(f_n, f_n) = (f_n, f_0) \\ t_1(f_0, f_1) + t_2(f_0, f_2) + \cdots + t_n(f_0, f_n) = (f_0, f_0) - d^2 \end{cases} \tag{6.10}$$

Excluding t_k from the system, we get $d^2 = \frac{\Gamma(f_0, f_1, f_2, \ldots, f_n)}{\Gamma(f_1, f_2, \ldots, f_n)}$. Formula (6.6) also follows from Remark 7.2. $\qquad\square$

Remark 6.1. From system (6.10), we conclude that $At = b$, where $b = (f_k, f_0)_{k=1}^n \in \mathbb{R}^n$, and hence $t = A^{-1}b$. By (6.8), we get

$$d^2 = (f_0, f_0) - (A^{-1}b, b) = \frac{\Gamma(f_0, f_1, f_2, \ldots, f_n)}{\Gamma(f_1, f_2, \ldots, f_n)}. \qquad (6.11)$$

See also Kosyak (2018, Chap. 4.3, Lemma 4.3.2).

6.3 Properties of m Infinite Vectors

The following statements hold for arbitrary m; see Kosyak (2025a, 2023).

Lemma 6.4 (Kosyak, 2025a). *Let f_0, f_1, f_2 be three infinite real vectors $f_r = (f_{rk})_{k\in\mathbb{N}}$, $0 \leq r \leq 2$. Then, for all r, s, where $0 \leq r < s \leq 2$ holds,*

$$\frac{\Gamma(f_0, f_1, f_2)}{\Gamma(f_r, f_s)} := \lim_{n\to\infty} \frac{\Gamma(f_0^{(n)}, f_1^{(n)}, f_2^{(n)})}{\Gamma(f_r^{(n)}, f_s^{(n)})} = \infty, \qquad (6.12)$$

if and only if for all $(C_0, C_1, C_2) \in \mathbb{R}^3 \setminus \{0\}$, it holds that $\sum_{r=0}^2 C_r f_r \notin l_2(\mathbb{N})$, and $C_r f_r + C_s f_s \notin l_2(\mathbb{N})$ for all $0 \leq r < s \leq 2$ and all $(C_r, C_s) \in \mathbb{R}^2 \setminus \{0\}$, where $f_r^{(n)} = (f_{rk})_{k=1}^n$.

Lemma 6.5 (Kosyak, 2025a, Lemma 3.4). *For $m = 3$, we have*

$$(C^{-1}(\lambda)a, a) = \Delta(y_1, y_2, y_3) \stackrel{(7.35)}{=} \frac{\det\big(I_3 + \gamma(y_1, y_2, y_3)\big)}{\det\big(I_2 + \gamma(y_2, y_3)\big)} - 1, \quad (6.13)$$

where $a = (a_{1k})_{k\in\mathbb{Z}}$, $y_r = (x_{rk})_{k\in\mathbb{Z}}$ are defined by (7.11) and $\lambda = (\lambda_k)_{k\in\mathbb{Z}}$.

Lemma 6.6 (Kosyak, 2025a). *Let $(y_k)_{k=1}^3$ be three real vectors such that $\sum_{k=1}^3 C_k y_k \notin l_2(\mathbb{Z})$ for any nontrivial combination*

$(C_k)_{k=1}^3 \in \mathbb{R}^3 \setminus \{0\}$. *Then,*

$$\frac{\det\left(I_3 + \gamma(y_1, y_2, y_2)\right)}{\det\left(I_2 + \gamma(y_2, y_3)\right)} = \lim_{n \to \infty} \frac{\det\left(I_3 + \gamma(y_1^{(n)}, y_2^{(n)}, y_3^{(n)})\right)}{\det\left(I_2 + \gamma(y_2^{(n)}, y_3^{(n)})\right)} = \infty, \quad (6.14)$$

where $y_r^{(n)} = (x_{rk})_{k=-n}^n \in \mathbb{R}^{2n+1}$.

Proof. The proof follows from Lemma 6.4 and (3.8). $\qquad\qquad\square$

6.4   Comparison of Two Gaussian Measures

For two centred Gaussian measures $\mu = \mu_{(b,0)}$ and $\mu' = \mu_{(b',0)}$ on the real line $\mathbb{R}$, defined by (3.5), it is well known that the *Hellinger integral* $H(\mu, \mu')$ is

$$H(\mu_{(b,0)}, \mu_{(b',0)}) = \left(\frac{4bb'}{(b+b')^2}\right)^{1/4}. \quad (6.15)$$

By Kakutani's criterion for product measures on $\mathbb{R}^{\mathbb{Z}}$ (Kakutani, 1948) and (6.15), we see that the following lemma holds true.

Lemma 6.7. *Two Gaussian measures* $\mu_{(b,0)}^1 = \otimes_{n \in \mathbb{Z}} \mu_{(b_n,0)}$ *and* $\mu_{(b',0)}^1 = \otimes_{n \in \mathbb{Z}} \mu_{(b'_n,0)}$, *where* $b = (b_n)_{n \in \mathbb{Z}}$ *and* $b' = (b'_n)_{n \in \mathbb{Z}}$, *are equivalent if and only if the product*

$$\prod_{n \in \mathbb{Z}} \frac{4b_n b'_n}{(b_n + b'_n)^2} \quad (6.16)$$

does not converge to 0. *An equivalent condition is*

$$\sum_{n \in \mathbb{Z}} \left(\sqrt{\frac{b_n}{b'_n}} - \sqrt{\frac{b'_n}{b_n}}\right)^2 < \infty. \quad (6.17)$$

Consider two measures: $\mu_{(\mathbb{I},0)} = \otimes_{n \in \mathbb{Z}} \mu_{(1,0)}$ and $\mu_{(\mathbb{I}+c,0)} = \otimes_{n \in \mathbb{Z}} \mu_{(1+c_n,0)}$ on the space X_1, where the measure $\mu_{(b,a)}$ on the real line $\mathbb{R}$ is defined by (3.5).

Lemma 6.8. *The two measures* $\mu_{(\mathbb{I},0)}$ *and* $\mu_{(\mathbb{I}+c,0)}$ *are equivalent if and only if* $\sum_{n \in \mathbb{Z}} c_n^2 < \infty$.

Proof. By Lemma 6.7 and (6.17),

$$\mu_{(\mathbb{I},0)} \sim \mu_{(\mathbb{I}+c,0)} \quad \Leftrightarrow \quad \sum_{n\in\mathbb{Z}} \left(\frac{1}{\sqrt{1+c_n}} - \sqrt{1+c_n} \right)^2 = \sum_{n\in\mathbb{Z}} \frac{c_n^2}{1+c_n} < \infty.$$

By Lemma 3.3, two series $\sum_{n\in\mathbb{Z}} \frac{c_n^2}{1+c_n}$ and $\sum_{n\in\mathbb{Z}} c_n^2$ are equivalent. $\quad\square$

The following lemma is also a consequence of Kakutani's criterion (Kakutani, 1948).

Lemma 6.9. *Two Gaussian measures $\mu_{(b,0)}^m$ and $\mu_{(b',0)}^m$ are equivalent if and only if the product*

$$\prod_{r=1}^{m} \prod_{n\in\mathbb{Z}} \frac{4b_{rn}b'_{rn}}{(b_{rn} + b'_{rn})^2} \tag{6.18}$$

does not converge to 0. The equivalent condition is

$$\sum_{r=1}^{m} \sum_{n\in\mathbb{Z}} \left(\sqrt{\frac{b_{rn}}{b'_{rn}}} - \sqrt{\frac{b'_{rn}}{b_{rn}}} \right)^2 < \infty. \tag{6.19}$$

Lemma 6.10 (Kosyak, 2004, 2018). *For $t\in \mathrm{GL}(m,\mathbb{R})\backslash\{e\}$, we have $(\mu_{(b,a)}^m)^{L_t} \perp \mu_{(b,a)}^m$ if and only if*

$$(\mu_{(b,0)}^m)^{L_t} \perp \mu_{(b,0)}^m \quad or \quad \mu_{(b,L_t a)}^m \perp \mu_{(b,a)}^m. \tag{6.20}$$

Chapter 7

The Generalised Characteristic Polynomial

7.1 Characteristic Polynomials

Consider an $n \times n$ matrix C. The *characteristic polynomial* of C, denoted by $p_C(t)$, is the polynomial defined by $p_C(t) = \det(tI - C)$, where I denotes the $n \times n$ identity matrix. By the *Cayley–Hamilton theorem*, we have $p_C(C) = 0$. Some authors define the characteristic polynomial as $p_C(t) = \det(C - tI)$. For a 2×2 matrix C, the characteristic polynomial is thus given by

$$p_C(t) = t^2 - t\operatorname{tr}(C) + \det C.$$

Using the language of *exterior algebras*, the characteristic polynomial of an $n \times n$ matrix C may be expressed as

$$p_C(t) = \sum_{k=0}^{n} t^{n-k}(-1)^k \operatorname{tr}\left(\wedge^k C\right) = \sum_{k=0}^{n} t^{n-k}(-1)^k c_k, \qquad (7.1)$$

where $c_k = \operatorname{tr}\left(\wedge^k C\right)$ is the trace of the kth *exterior power* of C, which has dimension $\binom{n}{k}$. This trace may be computed as *the sum of all principal minors of C of size k* (see Definition 7.2 and Remark 7.1):

$$c_k = \sum_{\emptyset \subseteq \alpha \subseteq \{1,2,\ldots,n\},\, |\alpha|=k} M_\alpha^\alpha(C). \qquad (7.2)$$

113

The recursive *Faddeev–LeVerrier algorithm* computes these coefficients more efficiently (Hou, 1998).

When the characteristic of the field of the coefficients is 0, each such trace may alternatively be computed as a single determinant, that of the $k \times k$ matrix:

$$
c_k = \operatorname{tr}\left(\bigwedge\nolimits^k C\right) = \frac{1}{k!}
\begin{vmatrix}
\operatorname{tr} C & k-1 & 0 & \cdots & 0 \\
\operatorname{tr} C^2 & \operatorname{tr} C & k-2 & \cdots & 0 \\
\vdots & \vdots & & \ddots & \vdots \\
\operatorname{tr} C^{k-1} & \operatorname{tr} C^{k-2} & & \cdots & 1 \\
\operatorname{tr} C^k & \operatorname{tr} C^{k-1} & & \cdots & \operatorname{tr} C
\end{vmatrix}. \tag{7.3}
$$

Theorem 7.1, formula (7.19), gives the expression for $\left(C(\lambda)\right)^{-1}$, where $C(\lambda) = C + \operatorname{diag}(\lambda_1, \ldots, \lambda_n)$. In particular, for the *resolvent* $(tI - C)^{-1}$, we have

$$
(tI - C)^{-1} = \frac{1}{p_C(t)} \left[\sum_{k=1}^{n} t^{n-k}(-1)^{k+1} \sum_{\alpha \subseteq \{1,2,\ldots,n\},\, |\alpha|=k} A^T(C_\alpha) \right], \tag{7.4}
$$

where the notations $A(C_\alpha)$ are defined in Definition 7.4. For a 3×3 matrix C, we have for example,

$$
(tI - C)^{-1} = \frac{1}{p_C(t)} \left[t^2 \sum_{k=1}^{3} A^T(C_k) - t \sum_{1 \le k < r \le 3} A^T(C_{kr}) + A^T(C_{123}) \right]. \tag{7.5}
$$

7.2   The Generalised Characteristic Polynomial

Definition 7.1. For a matrix $C \in \operatorname{Mat}(n, \mathbb{C})$ and $\lambda = (\lambda_k)_{k=1}^n \in \mathbb{C}^n$, define the *generalisation of the characteristic polynomial* $p_C(t) = \det(tI - C)$, $t \in \mathbb{C}$ as follows:

$$
P_C(\lambda) = \det C(\lambda), \quad \text{where} \quad C(\lambda) = \operatorname{diag}(\lambda_1, \ldots, \lambda_n) + C. \tag{7.6}
$$

Definition 7.2. For a matrix $C \in \mathrm{Mat}(n, \mathbb{C})$, $a \in \mathbb{C}^n$ and fixed $1 \le i_1 < i_2 < \cdots i_r \le n$ rows and $1 \le j_1 < j_2 < \cdots j_r \le n$ columns $1 \le r \le n$, denote by

$$M^{i_1 i_2 \ldots i_r}_{j_1 j_2 \ldots j_r}(C) \quad \text{and} \quad A^{i_1 i_2 \ldots i_r}_{j_1 j_2 \ldots j_r}(C)$$

the corresponding *minors* and *cofactors* of the matrix C.

Lemma 7.1 (Kosyak, 2018, Ch.1.4.3). *For the generalised characteristic polynomial $P_C(\lambda)$ of $C \in \mathrm{Mat}(n, \mathbb{C})$ and $\lambda = (\lambda_1, \lambda_2, \ldots, \lambda_n) \in \mathbb{C}^n$, we have*

$$P_C(\lambda) = \det C + \sum_{r=1}^{n} \sum_{1 \le i_1 < i_2 < \cdots < i_r \le n} \lambda_{i_1} \lambda_{i_2} \cdots \lambda_{i_r} A^{i_1 i_2 \cdots i_r}_{i_1 i_2 \cdots i_r}(C). \qquad (7.7)$$

Remark 7.1. If we set $\lambda_\alpha = \lambda_{i_1} \lambda_{i_2} \cdots \lambda_{i_r}$, where $\alpha = \{i_1, i_2, \ldots, i_r\}$ and $A^\alpha_\alpha(C) = A^{i_1 i_2 \cdots i_r}_{i_1 i_2 \cdots i_r}(C)$, $M^\alpha_\alpha(C) = M^{i_1 i_2 \cdots i_r}_{i_1 i_2 \cdots i_r}(C)$, $\lambda_\emptyset = 1$, $A^\emptyset_\emptyset(C) = \det C$ and $|\alpha| = r$, (see Definition 7.4), we may write (7.7) as follows:

$$P_C(\lambda) = \det C(\lambda) = \sum_{\emptyset \subseteq \alpha \subseteq \{1,2,\ldots,n\}} \lambda_\alpha A^\alpha_\alpha(C). \qquad (7.8)$$

Writing $\widehat{\alpha} = \{1, 2, \ldots, n\} \setminus \alpha$, we have $A^\alpha_\alpha(C) = M^{\widehat{\alpha}}_{\widehat{\alpha}}(C)$, hence

$$P_C(\lambda) = \det C(\lambda) = \left(\prod_{k=1}^{n} \lambda_k \right) \sum_{\emptyset \subseteq \alpha \subseteq \{1,2,\ldots,n\}} \frac{M^\alpha_\alpha(C)}{\lambda_\alpha}. \qquad (7.9)$$

7.3 Summary of the Key Formulas

For the generalised characteristic polynomial

$$P_C(\lambda) = \det \left(C + \mathrm{diag}(\lambda_1, \ldots, \lambda_n) \right),$$

we have the following (for notations, see Definitions 7.2, 7.4 and Remark 7.1):

$$P_C(\lambda) = \det C(\lambda) = \sum_{\emptyset \subseteq \alpha \subseteq \{1,2,\dots,n\}} \lambda_\alpha A_\alpha^\alpha(C)$$

$$= \left(\prod_{k=1}^n \lambda_k\right) \sum_{\emptyset \subseteq \alpha \subseteq \{1,2,\dots,n\}} \frac{M_\alpha^\alpha(C)}{\lambda_\alpha},$$

$$C(\lambda)^{-1} = \frac{1}{P_C(\lambda)} \left(\prod_{k=1}^n \lambda_k\right) \sum_{\emptyset \neq \alpha \subseteq \{1,2,\dots,n\}} \frac{A^T(C_\alpha)}{\lambda_\alpha},$$

$$(C(\lambda)^{-1}a, a) = \frac{1}{P_C(\lambda)} \left(\prod_{k=1}^n \lambda_k\right) \sum_{\emptyset \neq \alpha \subseteq \{1,2,\dots,n\}} \frac{(A^T(C_\alpha)a_\alpha, a_\alpha)}{\lambda_\alpha},$$

$$1 + (C(\lambda)^{-1}a, a) = \frac{\det(C(\lambda) + a \otimes a)}{\det C(\lambda)}, \quad \text{where} \quad a \otimes a = (a_k a_r)_{k,r=1}^n.$$

Another representation of $1 + (C(\lambda)^{-1}a, a)$, which we will use, is given in Theorem 7.2.

7.4   Gram Determinants and Gram Matrices

Definition 7.3. Gram determinants were introduced in 1879 by Gram. For vectors $x_1, x_2, \dots, x_n$ in some Hilbert space H, the *Gram matrix* $\gamma(x_1, x_2, \dots, x_n)$ is defined by the formula (see also Gantmacher, 1958, Chap IX, §5)

$$\gamma(x_1, x_2, \dots, x_n) = (x_k, x_m)_{k,m=1}^n.$$

The determinant of this matrix is called the *Gram determinant* for the vectors $x_1, x_2, \dots, x_n$ and is denoted by $\Gamma(x_1, x_2, \dots, x_n)$:

$$\Gamma(x_1, x_2, \dots, x_n) := \det \gamma(x_1, x_2, \dots, x_n). \tag{7.10}$$

Some authors use the notation $G(x_1, x_2, \dots, x_n)$.

Remark 7.2. A Gram determinant is equal to the square of the n−dimensional volume of the *parallelotope* constructed on $x_1, x_2, \ldots, x_n$.

Fix some notations:

$$X = X_{mn} = \begin{pmatrix} x_{11} & x_{12} & \cdots & x_{1n} \\ x_{21} & x_{22} & \cdots & x_{2n} \\ \vdots & \vdots & \ddots & \vdots \\ x_{m1} & x_{m2} & \cdots & x_{mn} \end{pmatrix}, \tag{7.11}$$

$$x_k = (x_{1k}, x_{2k}, \ldots, x_{mk}) \in \mathbb{R}^m, y_r = (x_{r1}, x_{r2}, \ldots, x_{rn}) \in \mathbb{R}^n. \tag{7.12}$$

Then, obviously, we get

$$X^*X = \begin{pmatrix} (x_1, x_1) & (x_1, x_2) & \cdots & (x_1, x_n) \\ (x_2, x_1) & (x_2, x_2) & \cdots & (x_2, x_n) \\ \vdots & \vdots & \ddots & \vdots \\ (x_n, x_1) & (x_n, x_2) & \cdots & (x_n, x_n) \end{pmatrix} = \gamma(x_1, x_2, \ldots, x_n), \tag{7.13}$$

$$XX^* = \begin{pmatrix} (y_1, y_1) & (y_1, y_2) & \cdots & (y_1, y_m) \\ (y_2, y_1) & (y_2, y_2) & \cdots & (y_2, y_m) \\ \vdots & \vdots & \ddots & \vdots \\ (y_m, y_1) & (y_m, y_2) & \cdots & (y_m, y_m) \end{pmatrix} = \gamma(y_1, y_2, \ldots, y_m). \tag{7.14}$$

Therefore, we have

$$\Gamma(x_1, x_2, \ldots, x_n) = \det(X^*X) = \det(XX^*) = \Gamma(y_1, y_2, \ldots, y_m). \tag{7.15}$$

7.5 The Explicit Expression for $C^{-1}(\lambda)$ and $(C^{-1}(\lambda)a, a)$

Fix $C \in \mathrm{Mat}(n, \mathbb{C})$, $a \in \mathbb{C}^n$ and $\lambda \in \mathbb{C}^n$. Our aim is to find the explicit formulas for $C(\lambda)^{-1}$ and $(C(\lambda)^{-1}a, a)$, where $C(\lambda)$ is defined by (7.6).

Definition 7.4. For $\alpha = \{i_1, i_2, \ldots, i_r\} \subset \{1, 2, \ldots, n\}$, set $M(\alpha)(C) = M_\alpha^\alpha(C)$. Let also $C_\alpha = C_{i_1 i_2 \ldots i_r}$ be the corresponding *submatrix* of the matrix C and $a_\alpha = (a_{i_1}, a_{i_2}, \ldots, a_{i_r})$. The elements of the matrix C_α are located at the intersection of the $i_1, i_2, \ldots, i_r$ rows and columns of the matrix C. Denote by $A(C_{i_1 i_2 \ldots i_r})$ the matrix of the cofactors of the first order of the matrix $C_{i_1 i_2 \ldots i_r}$, which is also known as the *adjugate matrix* or *adjunct matrix*:

$$A(C_{i_1 i_2 \ldots i_r}) = (A_j^i(C_{i_1 i_2 \ldots i_r}))_{1 \le i,j \le r}. \tag{7.16}$$

The minor of order zero is often defined to be 1; therefore, we set $A(C_k) = 1$ for $1 \le k \le n$. As usual, denote by B^T the *matrix transposed* to B.

Let $n = 3$. Then, $A(C_{123}) = A(C)$ is the following matrix:

$$A(C) = A(C_{123}) = \begin{pmatrix} A_1^1 & A_2^1 & A_3^1 \\ A_1^2 & A_2^2 & A_3^2 \\ A_1^3 & A_2^3 & A_3^3 \end{pmatrix} = \begin{pmatrix} M_{23}^{23} & -M_{13}^{23} & M_{12}^{23} \\ -M_{23}^{13} & M_{13}^{13} & -M_{12}^{13} \\ M_{23}^{12} & -M_{13}^{12} & M_{12}^{12} \end{pmatrix},$$
$$\tag{7.17}$$

where we write M_{rs}^{ij} instead of $M_{rs}^{ij}(C)$ and A_j^i instead of $A_j^i(C)$.

Remark 7.3. If $\det C_{i_1 i_2 \ldots i_r} \ne 0$, we have

$$A^T(C_{i_1 i_2 \ldots i_r}) = \det C_{i_1 i_2 \ldots i_r}(C_{i_1 i_2 \ldots i_r})^{-1}. \tag{7.18}$$

In what follows, we need to consider the submatrix $A^T(C_{i_1 i_2 \ldots i_r})$, $1 \le r \le n$, of the matrix $C \in \mathrm{Mat}(n, \mathbb{C})$ as an *appropriate element of* $\mathrm{Mat}(n, \mathbb{C})$.

Theorem 7.1. *For the matrix $C(\lambda)$, defined by* (7.6), $a \in \mathbb{C}^n$ *and* $\lambda \in \mathbb{C}^n$, *we have*

$$C(\lambda)^{-1} = \frac{1}{P_C(\lambda)} \left(\prod_{k=1}^n \lambda_k \right) \sum_{r=1}^n \sum_{1 \le i_1 < i_2 < \ldots i_r \le n} \frac{A^T(C_{i_1 i_2 \ldots i_r})}{\lambda_{i_1} \lambda_{i_2} \ldots \lambda_{i_r}},$$
$$\tag{7.19}$$

$$(C(\lambda)^{-1}a,a) = \frac{1}{P_C(\lambda)} \left(\prod_{k=1}^{n} \lambda_k\right) \sum_{r=1}^{n} \sum_{1\leq i_1<i_2<\cdots<i_r\leq n}$$

$$\times \frac{(A^T(C_{i_1i_2\ldots i_r})a_{i_1i_2\ldots i_r}, a_{i_1i_2\ldots i_r})}{\lambda_{i_1}\lambda_{i_2}\ldots\lambda_{i_r}}, \tag{7.20}$$

$$(C(\lambda)^{-1}a,a) = \frac{1}{P_C(\lambda)} \left(\prod_{k=1}^{n} \lambda_k\right) \sum_{\alpha\subseteq\{1,2,\ldots,n\},\,|\alpha|\geq 1} \frac{(A^T(C_\alpha)a_\alpha, a_\alpha)}{\lambda_\alpha}. \tag{7.21}$$

Proof. For $n=2$, we have

$$C(\lambda) = \begin{pmatrix} c_{11}+\lambda_1 & c_{12} \\ c_{21} & c_{22}+\lambda_2 \end{pmatrix}, \quad A^T(C_{12}) = \begin{pmatrix} A_1^1 & A_1^2 \\ A_2^1 & A_2^2 \end{pmatrix} = \begin{pmatrix} c_{22} & -c_{12} \\ -c_{21} & c_{11} \end{pmatrix}, \tag{7.22}$$

$$C(\lambda)^{-1} = \frac{1}{P_C(\lambda)} \begin{pmatrix} c_{22}+\lambda_2 & -c_{12} \\ -c_{21} & c_{11}+\lambda_1 \end{pmatrix} = \frac{1}{P_C(\lambda)} \left[\begin{pmatrix} \lambda_2 & 0 \\ 0 & \lambda_1 \end{pmatrix} + A^T(C_{12}) \right]$$

$$= \frac{\lambda_1\lambda_2}{P_C(\lambda)} \left[\begin{pmatrix} \lambda_1^{-1} & 0 \\ 0 & \lambda_2^{-1} \end{pmatrix} + \frac{A^T(C_{12})}{\lambda_1\lambda_2} \right]$$

$$= \frac{\lambda_1\lambda_2}{P_C(\lambda)} \left[\sum_{k=1}^{2} \frac{A^T(C_k)}{\lambda_k} + \frac{A^T(C_{12})}{\lambda_1\lambda_2} \right], \tag{7.23}$$

recalling that $A^T(C_1) = \left(\begin{smallmatrix} 1 & 0 \\ 0 & 0 \end{smallmatrix}\right)$, $A^T(C_2) = \left(\begin{smallmatrix} 0 & 0 \\ 0 & 1 \end{smallmatrix}\right)$. Therefore,

$$(C(\lambda)^{-1}a,a) = \frac{1}{P_C(\lambda)} \left[(c_{22}+\lambda_2)a_1^2 - (c_{12}+c_{21})a_1a_2 + (c_{11}+\lambda_1)a_2^2 \right]$$

$$= \frac{1}{P_C(\lambda)} \left[\lambda_2 a_1^2 + \lambda_1 a_2^2 + c_{22}a_1^2 + c_{11}a_2^2 - (c_{12}+c_{21})a_1a_2 \right]$$

$$= \left(1 + \frac{M(1)}{\lambda_1} + \frac{M(2)}{\lambda_2} + \frac{M(12)}{\lambda_1\lambda_2} \right)^{-1}$$

$$\times \left[\frac{a_1^2}{\lambda_1} + \frac{a_2^2}{\lambda_2} + \frac{(A^T(C_{12})a_{12}, a_{12})}{\lambda_1\lambda_2} \right]. \tag{7.24}$$

For $n = 3$, we have by (7.9),

$$P_C(\lambda) = \lambda_1 \lambda_2 \lambda_3 \left(1 + \sum_{k=1}^{3} \frac{M(k)}{\lambda_k} + \sum_{1 \le k < r \le 3} \frac{M(kr)}{\lambda_k \lambda_r} + \frac{M(123)}{\lambda_1 \lambda_2 \lambda_3} \right),$$

$$(7.25)$$

$$C(\lambda) = \begin{pmatrix} c_{11} + \lambda_1 & c_{12} & c_{13} \\ c_{21} & c_{22} + \lambda_2 & c_{23} \\ c_{31} & c_{32} & c_{33} + \lambda_3 \end{pmatrix},$$

$$C(\lambda)^{-1} = \frac{1}{P_C(\lambda)} \begin{pmatrix} \lambda_2 \lambda_3 \left(1 + \frac{M_2^2}{\lambda_2} + \frac{M_3^3}{\lambda_3} + \frac{M_{23}^{23}}{\lambda_2 \lambda_3} \right) & -M_{23}^{13} - \lambda_3 M_2^1 & M_{23}^{12} - \lambda_2 M_3^1 \\ -M_{13}^{23} - \lambda_3 M_1^2 & \lambda_1 \lambda_3 \left(1 + \frac{M_1^1}{\lambda_1} + \frac{M_3^3}{\lambda_3} + \frac{M_{13}^{13}}{\lambda_1 \lambda_3} \right) & -M_{13}^{12} - \lambda_1 M_3^2 \\ -M_{12}^{23} - \lambda_2 M_1^3 & -M_{12}^{13} - \lambda_1 M_2^3 & \lambda_1 \lambda_2 \left(1 + \frac{M_1^1}{\lambda_1} + \frac{M_2^2}{\lambda_2} + \frac{M_{12}^{12}}{\lambda_1 \lambda_2} \right) \end{pmatrix}$$

$$= \frac{\lambda_1 \lambda_2 \lambda_3}{P_C(\lambda)} \begin{pmatrix} \frac{1}{\lambda_1} + \frac{M_2^2}{\lambda_1 \lambda_2} + \frac{M_3^3}{\lambda_1 \lambda_3} + \frac{M_{23}^{23}}{\lambda_1 \lambda_2 \lambda_3} & -\frac{M_{23}^{13}}{\lambda_1 \lambda_2 \lambda_3} - \frac{M_2^1}{\lambda_1 \lambda_2} & \frac{M_{23}^{12}}{\lambda_1 \lambda_2 \lambda_3} - \frac{M_3^1}{\lambda_1 \lambda_3} \\ -\frac{M_{13}^{23}}{\lambda_1 \lambda_2 \lambda_3} - \frac{M_1^2}{\lambda_1 \lambda_2} & \frac{1}{\lambda_2} + \frac{M_1^1}{\lambda_1 \lambda_2} + \frac{M_3^3}{\lambda_2 \lambda_3} + \frac{M_{13}^{13}}{\lambda_1 \lambda_2 \lambda_3} & -\frac{M_{13}^{12}}{\lambda_1 \lambda_2 \lambda_3} - \frac{M_3^2}{\lambda_2 \lambda_3} \\ -\frac{M_{12}^{23}}{\lambda_1 \lambda_2 \lambda_3} - \frac{M_1^3}{\lambda_1 \lambda_3} & -\frac{M_{12}^{13}}{\lambda_1 \lambda_2 \lambda_3} - \frac{M_2^3}{\lambda_2 \lambda_3} & \frac{1}{\lambda_3} + \frac{M_1^1}{\lambda_1 \lambda_3} + \frac{M_2^2}{\lambda_2 \lambda_3} + \frac{M_{12}^{12}}{\lambda_1 \lambda_2 \lambda_3} \end{pmatrix}.$$

Finally, we get

$$C(\lambda)^{-1} = \frac{\lambda_1 \lambda_2 \lambda_3}{P_C(\lambda)} \left[\sum_{k=1}^{3} \frac{A^T(C_k)}{\lambda_k} + \sum_{1 \le r < s \le 3} \frac{A^T(C_{rs})}{\lambda_r \lambda_s} + \frac{A^T(C_{123})}{\lambda_1 \lambda_2 \lambda_3} \right],$$

$$(7.26)$$

and we use (7.22) and (7.17). Therefore,

$$(C(\lambda)^{-1}a, a) = \frac{\lambda_1\lambda_2\lambda_3}{P_C(\lambda)}\left[\frac{a_1^2}{\lambda_1} + \frac{a_2^2}{\lambda_2} + \frac{a_3^2}{\lambda_3} + \frac{(A^T(C_{12})a_{12}, a_{12})}{\lambda_1\lambda_2}\right.$$

$$\left. + \frac{(A^T(C_{13})a_{13}, a_{13})}{\lambda_1\lambda_3} + \frac{(A^T(C_{23})a_{23}, a_{23})}{\lambda_2\lambda_3} + \frac{(A^T(C_{123})a_{123}, a_{123})}{\lambda_1\lambda_2\lambda_3}\right].$$

$$(7.27)$$

For $n = 4$, we have

$$C(\lambda) = \begin{pmatrix} c_{11} + \lambda_1 & c_{12} & c_{13} & c_{14} \\ c_{21} & c_{22} + \lambda_2 & c_{23} & c_{24} \\ c_{31} & c_{32} & c_{33} + \lambda_3 & c_{34} \\ c_{41} & c_{42} & c_{43} & c_{44} + \lambda_4 \end{pmatrix}.$$

The general formulas are as follows:

$$C(\lambda)^{-1} = \frac{1}{P_C(\lambda)}\left(\prod_{k=1}^{n}\lambda_k\right)\sum_{\alpha\subseteq\{1,2,\ldots,n\},\,|\alpha|\geq 1}\frac{A^T(C_\alpha)}{\lambda_\alpha},$$

$$(C(\lambda)^{-1}a, a) = \frac{1}{P_C(\lambda)}\left(\prod_{k=1}^{n}\lambda_k\right)\sum_{\alpha\subseteq\{1,2,\ldots,n\},\,|\alpha|\geq 1}\frac{(A^T(C_\alpha)a_\alpha, a_\alpha)}{\lambda_\alpha}.$$

This proves (7.19)–(7.21). We make use of a convention in (7.20) that $A(C_k) = 1$. $\qquad\square$

Example 7.1. For the matrix $C(\lambda)$, we have by (7.9),

$$C(\lambda) = \begin{pmatrix} 1 + \lambda_1 & 1 & \cdots & 1 \\ 1 & 1 + \lambda_2 & \cdots & 1 \\ & & \vdots & \\ 1 & 1 & \cdots & 1 + \lambda_n \end{pmatrix}, \qquad (7.28)$$

$$\det C(\lambda) = \left(\prod_{k=1}^{n} \lambda_k \right) \left(1 + \sum_{k=1}^{n} \frac{1}{\lambda_k} \right), \tag{7.29}$$

$$C(\lambda)^{-1} = \left(1 + \sum_{k=1}^{n} \frac{1}{\lambda_k} \right)^{-1} \left[\sum_{k=1}^{n} \frac{A^T(C_k)}{\lambda_k} + \sum_{1 \le k < r \le n} \frac{A^T(C_{kr})}{\lambda_k \lambda_r} \right], \tag{7.30}$$

where $A^T(C_{kr}) = \left(\begin{smallmatrix} 1 & -1 \\ -1 & 1 \end{smallmatrix} \right)$ and $A^T(C_{krs}) = 0$ for $1 \le k < r < s \le n$.

7.6   The Case Where C Is the Gram Matrix

Fix two natural numbers $n, m \in \mathbb{N}$ with $m \le n$, two matrices A_{mn} and X_{mn}, vectors $g_k \in \mathbb{C}^{m-1}$, $1 \le k \le n$ and $a \in \mathbb{C}^n$ as follows:

$$A_{mn} = \begin{pmatrix} a_{11} & a_{12} & \cdots & a_{1n} \\ a_{21} & a_{22} & \cdots & a_{2n} \\ & & \vdots & \\ a_{m1} & a_{m2} & \cdots & a_{mn} \end{pmatrix}, \quad g_k = \begin{pmatrix} a_{2k} \\ a_{3k} \\ \vdots \\ a_{mk} \end{pmatrix} \in \mathbb{C}^{m-1},$$

$$a = (a_{1k})_{k=1}^{n} \in \mathbb{C}^n. \tag{7.31}$$

Set

$$C = \gamma(g_1, g_2, \ldots, g_n) = \begin{pmatrix} (g_1, g_1) & (g_1, g_2) & \cdots & (g_1, g_n) \\ (g_2, g_1) & (g_2, g_2) & \cdots & (g_2, g_n) \\ & & \vdots & \\ (g_n, g_1) & (g_n, g_2) & \cdots & (g_n, g_n) \end{pmatrix}. \tag{7.32}$$

We calculate $P_C(\lambda)$, $C^{-1}(\lambda)$ and $(C^{-1}(\lambda)a, a)$ for an arbitrary n. Consider the matrix

$$X_{mn} = \begin{pmatrix} x_{11} & x_{12} & \cdots & x_{1n} \\ x_{21} & x_{22} & \cdots & x_{2n} \\ & & \vdots & \\ x_{m1} & x_{m2} & \cdots & x_{mn} \end{pmatrix}, \quad \text{where} \quad x_{rk} = \frac{a_{rk}}{\sqrt{\lambda_k}}, \tag{7.33}$$

$$\bar{x}_k = (x_{rk})_{r=2}^{m} = \frac{g_k}{\sqrt{\lambda_k}} \in \mathbb{C}^{m-1}. \tag{7.34}$$

For $k \in \mathbb{N}$, define $\Delta(y_1, y_2, \ldots, y_k)$ as follows:

$$\Delta(y_1, y_2, \ldots, y_k) = \frac{\det(I + \gamma(y_1, y_2, \ldots, y_k))}{\det(I + \gamma(y_2, \ldots, y_k))} - 1. \qquad (7.35)$$

Lemma 7.2. *For $A \in \mathrm{GL}(n, \mathbb{C})$ and $a \in \mathbb{C}^n$, we have*

$$1 + (A^{-1}a, a) = \frac{\det(A + a \otimes a)}{\det(A)}. \qquad (7.36)$$

Proof. This is a variant of the Shermann–Morrison theorem.[1] Define $a \otimes a$ as $(a_k a_r)_{k,r=1}^n \in \mathrm{Mat}(n, \mathbb{C})$. Then, by (7.1), we have

$$\det(A + a \otimes a) = \det(A)\det(1 + A^{-1}a \otimes a) = \det(A)(1 + (A^{-1}a, a)),$$

since $\mathrm{tr}(A^{-1}a \otimes a) = (A^{-1}a, a)$ and $c_k = \mathrm{tr}(\wedge^k(A^{-1}a \otimes a)) = 0$ for all $k > 1$. To verify the last statement, by (7.3), it is sufficient to verify that $\mathrm{tr}(D^k) = (\mathrm{tr}\, D)^k$ for $D = A^{-1}a \otimes a$. Indeed, we have $\mathrm{tr}\, D = (A^{-1}a, a)$ and

$$D^k = (A^{-1}a, a)^{k-1}D, \quad \text{therefore,} \quad \mathrm{tr}(D^k) = (\mathrm{tr}\, D)^k. \qquad (7.37)$$

$\square$

If we take $A = C(\lambda)$, we would get

$$1 + (C(\lambda)^{-1}a, a) = \frac{\det(C(\lambda) + a \otimes a)}{\det C(\lambda)}. \qquad (7.38)$$

Theorem 7.2. *Let C be defined by (7.32) and $a, \lambda \in \mathbb{C}^n$. Then,*

$$1 + (C(\lambda)^{-1}a, a) = \frac{\det(I_m + \gamma(y_1, y_2, \ldots, y_m))}{\det(I_{m-1} + \gamma(y_2, \ldots, y_m))}$$
$$= 1 + \Delta(y_1, y_2, \ldots, y_m), \qquad (7.39)$$

where y_k for $1 \leq k \leq m$ are defined by (7.12) and $\Delta(y_1, y_2, \ldots, y_m)$ is defined by (7.35).

---

[1]We thank Dr O. Gamayun from the London Institute for Mathematical Sciences for drawing our attention to this theorem.

Proof. By Lemma 7.2, it is sufficient to show that

$$\det\big(C(\lambda) + a \otimes a\big) = \left(\prod_{k=1}^{n} \lambda_k\right)\det(I + \gamma(y_1, \ldots, y_m)),$$

$$\det C(\lambda) = \left(\prod_{k=1}^{n} \lambda_k\right)\det\big(I+\gamma(y_2, \ldots, y_m)\big).$$

Indeed, we have

$$
\begin{aligned}
C(\lambda) + a \otimes a &= \gamma(g_1, \ldots, g_n) + \operatorname{diag}(\lambda_k)_{k=1}^{n} + (a_{1k}a_{1r})_{k,r=1}^{n} \\
&= \big((g_k, g_r) + a_{1k}a_{1r}\big)_{k,r=1}^{n} + \operatorname{diag}(\lambda_k)_{k=1}^{n} \\
&\overset{(7.33)}{=} \big((x_k, x_r)\sqrt{\lambda_k \lambda_r}\big)_{k,r=1}^{n} + \operatorname{diag}(\lambda_k)_{k=1}^{n} \\
&= \operatorname{diag}(\sqrt{\lambda_k})_{k=1}^{n}\Big(I + \gamma(x_1, \ldots, x_n)\Big)\operatorname{diag}(\sqrt{\lambda_k})_{k=1}^{n}.
\end{aligned}
$$

Therefore,

$$
\begin{aligned}
\det\big(C(\lambda) + a \otimes a\big) &= \left(\prod_{k=1}^{n} \lambda_k\right)\det\big(I + \gamma(x_1, \ldots, x_n)\big) \\
&\overset{(7.15)}{=} \left(\prod_{k=1}^{n} \lambda_k\right)\det\big(I + \gamma(y_1, \ldots, y_m)\big).
\end{aligned}
$$

Further,

$$
\begin{aligned}
\det C(\lambda) &= \det\Big(\gamma(g_1, \ldots, g_n) + \operatorname{diag}(\lambda_k)_{k=1}^{n}\Big) \\
&\overset{(7.34)}{=} \det\Big(\operatorname{diag}(\sqrt{\lambda_k})_{k=1}^{n}(I + \gamma(\bar{x}_1, \ldots, \bar{x}_n))\operatorname{diag}(\sqrt{\lambda_k})_{k=1}^{n}\Big) \\
&= \left(\prod_{k=1}^{n} \lambda_k\right)\det\big(I + \gamma(\bar{x}_1, \ldots, \bar{x}_n)\big) \\
&\overset{(7.15)}{=} \left(\prod_{k=1}^{n} \lambda_k\right)\det\big(I + \gamma(y_2, \ldots, y_m)\big).
\end{aligned}
$$

$\square$

Chapter 8

The Height of an Infinite Parallelotope

8.1 Introduction

We show that if no non-trivial linear combinations of vectors $f_0, f_1, \ldots, f_m \in \mathbb{R}^\infty$ belong to $l_2(\mathbb{N})$, then all the heights of an infinite parallelotope constructed on the vectors $f_0, f_1, \ldots, f_m$ are infinite. This result is essential to the proof of the irreducibility of unitary representations of some infinite-dimensional groups.

This chapter lies at the interface of linear algebra and functional analysis since we study the behaviour of a finite-dimensional object, namely a parallelotope generated by $m + 1$ vectors, belonging to an n-dimensional space of dimension $n \to \infty$. The parallelotope is a set which is made up of vectors $\sum C_i f_i$, where $C_i \in [0, 1]$. The main result is proved in Section 8.4. To solve Problem 8.1, in some particular cases, we use certain results from Section 6.2, namely Lemmas 6.1 and 6.3. They are the law of large numbers and some generalisations; see Section 8.1.1. A sketch of the proof of irreducibility is given after Problem 8.1. To prove irreducibility for some infinite-dimensional groups (Kosyak, 2018), we need to approximate a rich space of functions by combinations of generators of one-parameter subgroups. The desired approximation on the nth step is of the order of the inverse height h_n of the projection of the vectors f_k and the parallelotope to some finite-dimensional space $\mathbb{R}^n$. The result on the heights of an infinite parallelotope allows us to prove the approximation and irreducibility. In the proof, we also used the explicit formulas obtained in Kosyak (2018, 2023) for $\det C(\lambda)$, $C^{-1}(\lambda)$ and $(C^{-1}(\lambda)a, a)$, where $C(\lambda) = C + \operatorname{diag}(\lambda_1, \ldots, \lambda_n)$ and C is an $n \times n$ matrix.

Problem 8.1. We start with a general problem; see Section 8.1.1 later. For a fixed vector f_0 in a Hilbert space H and an infinite sequence of vectors $(f_n)_{n \in \mathbb{N}}$ in H, when

$$f_0 \in V = \langle f_n, n \in \mathbb{N} \rangle? \tag{8.1}$$

Here, V is the completion of all finite linear combinations of vectors from a family $(f_n)_{n \in \mathbb{N}}$. Let us denote by V_n the subspace generated by the first n vectors. Then, the square of the distance $d^2(f_0, V_n)$ of the vector f_0 from the hyperplane V_n is given by the ratio of two Gram determinants; see (6.6) above. Finally, $f_0 \in V$ if and only if $\lim_{n \to \infty} d^2(f_0, V_n) = 0$.

By Lemma 6.1, for a strictly positive operator A acting in $\mathbb{R}^n$ and a vector $b \in \mathbb{R}^n \backslash \{0\}$, we have

$$\min_{x \in \mathbb{R}^n} \left((Ax, x) \mid (x, b) = 1 \right) = \frac{1}{(A^{-1}b, b)}.$$

In the concrete examples considered in Kosyak (1992)–Kosyak (2023), the possibility of approximating a number of functions in $L^\infty(X_m, \mu)$ using Lemma 6.1 follows from the fact that

$$\lim_{n \to \infty} (C_n(\lambda)^{-1}a_n, a_n) = \infty. \tag{8.2}$$

Here, $C_n = \gamma(g_1, g_2, \ldots, g_n)$, where $\gamma(g_1, g_2, \ldots, g_n)$ is a Gram matrix of the vectors $g_1, g_2, \ldots, g_n \in H$ (see Definition 7.3) and $C_n(\lambda) = \operatorname{diag}(\lambda_1, \ldots, \lambda_n) + C_n$, $\lambda \in \mathbb{C}^n$. By Theorem 5.2 proved in Kosyak (2023), we have

$$(C_n(\lambda)^{-1}a_n, a_n) = \frac{\det\left(I_m + \gamma(y_1^{(n)}, y_2^{(n)}, \ldots, y_m^{(n)})\right)}{\det\left(I_{m-1} + \gamma(y_2^{(n)}, \ldots, y_m^{(n)})\right)} - 1. \tag{8.3}$$

Finally, by Lemma 8.4, we have

$$\lim_{n \to \infty} \frac{\det\big(I_m + \gamma(y_1^{(n)}, y_2^{(n)}, \ldots, y_m^{(n)})\big)}{\det\big(I_{m-1} + \gamma(y_2^{(n)}, \ldots, y_m^{(n)})\big)} = \infty.$$

8.1.1 *The law of large numbers*

Consider $\mathbb{R}^\infty$ with the infinite product of a *standard Gaussian measure*

$$\mu_1(x) = \bigotimes_{n=1}^{\infty} \mu(x_n), \quad \text{where} \quad d\mu(x_n) = \sqrt{\frac{1}{2\pi}} \exp\left(-\frac{x_n^2}{2}\right) dx_n. \quad (8.4)$$

Set $f_0(x) \equiv 1$ and $f_n = x_n^2$. The law of large numbers can be reformulated in this particular case as follows:

Lemma 8.1. *We have $f_0 \in \langle f_n, n \in \mathbb{N}\rangle$; moreover, in $H = L_2(\mathbb{R}^\infty, \mu_1)$, it holds that*

$$s.\lim_{n} \frac{1}{n} \sum_{k=1}^{n} x_k^2 = f_0. \quad (8.5)$$

Proof. To prove this lemma, we should show that $\lim_{n \to \infty} d^2(f_0, V_n) = 0$, where $d^2(f_0, V_n)$ is defined by (6.6). We should calculate $\Gamma(f_0, f_1, f_2, \ldots, f_n)$ and $\Gamma(f_1, f_2, \ldots, f_n)$. For the corresponding *Gram matrices* $\gamma(f_1, f_2, \ldots, f_n)$, we get (where we denote $\lambda_k = 2$ for all $1 \le k \le n$)

$$\gamma(f_0, f_1, f_2, \ldots, f_n) = \big((f_i, f_j)\big)_{i,j=0}^{n} = \begin{pmatrix} 1 & 1 & 1 \cdots & 1 \\ 1 & 3 & 1 \cdots & 1 \\ 1 & 1 & 3 \cdots & 1 \\ & & \vdots & \\ 1 & 1 & 1 \cdots & 3 \end{pmatrix}$$

$$= \begin{pmatrix} 1 & 1 & 1\cdots & & 1 \\ 1 & 1+\lambda_1 & 1\cdots & & 1 \\ 1 & 1 & 1+\lambda_2\cdots & & 1 \\ & & & \vdots & \\ 1 & 1 & 1\cdots & & 1+\lambda_n \end{pmatrix},$$

$$\gamma(f_1,f_2,\ldots,f_n) = \left((f_i,f_j)\right)_{i,j=1}^n = \begin{pmatrix} 3 & 1\cdots & 1 \\ 1 & 3\cdots & 1 \\ & & \vdots \\ 1 & 1\cdots & 3 \end{pmatrix}$$

$$= \begin{pmatrix} 1+\lambda_1 & 1\cdots & & 1 \\ 1 & 1+\lambda_2\cdots & & 1 \\ & & \vdots & \\ 1 & 1\cdots & & 1+\lambda_n \end{pmatrix},$$

$$\Gamma(f_0,f_1,f_2,\ldots,f_n) = \det\gamma(f_0,f_1,f_2,\ldots,f_n) \stackrel{(7.29)}{=} \prod_{k=1}^n \lambda_k,$$

$$\Gamma(f_0,f_1,f_2,\ldots,f_n)$$

$$= \det\gamma(f_0,f_1,f_2,\ldots,f_n) \stackrel{(7.29)}{=} \prod_{k=1}^n \lambda_k\left(1+\sum_{k=1}^n \frac{1}{\lambda_k}\right),$$

$$d^2(f_0,V_n) = \frac{\det\left(\gamma(f_0,f_1,\ldots,f_n)\right)}{\det\left(\gamma(f_1,f_2,\ldots,f_n)\right)} = \left(1+\sum_{k=1}^n \frac{1}{2}\right)^{-1} \to 0.$$

By Lemma 6.1 and (7.30), we conclude that at step n, the coefficient should be $t_k \equiv \frac{1}{n}$. $\qquad\square$

8.1.2 *Some generalisations*

Consider $\mathbb{R}^\infty \times \mathbb{R}^\infty$ with the infinite product of a standard Gaussian measure

$$\mu_2(x) = \bigotimes_{k=1}^{2} \bigotimes_{n=1}^{\infty} \mu(x_{kn}). \tag{8.6}$$

Set $f_0(x) \equiv 1$, $f_n = x_{1n}^2 + a_n x_{1n} x_{2n}$, where $a_n \neq 0$ for all $n \in \mathbb{N}$.

Lemma 8.2 (Kosyak, 2023). *We have $f_0 \in \langle f_n, n \in \mathbb{N} \rangle$ if and only if $\sum_{k=3}^{\infty} \frac{1}{a_k^2} = \infty$.*

Proof. To prove this lemma, we should show that $\lim_{n \to \infty} d^2(f_0, V_n) = 0$, where $d^2(f_0, V_n)$ is defined by (6.6), if and only if $\sum_{k=3}^{\infty} \frac{1}{a_k^2} = \infty$. Further,

$$\gamma(f_0, f_3, f_4, \ldots, f_{n+2}) = \begin{pmatrix} 1 & 1 & \cdots & 1 \\ 1 & 1+a_3^2 & \cdots & 1 \\ & & \vdots & \\ 1 & 1 & \cdots & 1+a_{n+2}^2 \end{pmatrix}, \tag{8.7}$$

$$\gamma(f_3, f_4, \ldots, f_{n+2}) = \begin{pmatrix} 1+a_3^2 & 1 & \cdots & 1 \\ 1 & 1+a_4^2 & \cdots & 1 \\ & & \vdots & \\ 1 & 1 & \cdots & 1+a_{n+2}^2 \end{pmatrix}, \tag{8.8}$$

$$\det\big(\gamma(f_0, f_3, f_4, \ldots, f_{n+2})\big) \overset{(7.29)}{=} \left(\prod_{k=3}^{n+2} a_k^2 \right), \tag{8.9}$$

$$\det\big(\gamma(f_3, f_4, \ldots, f_{n+2})\big) \overset{(7.29)}{=} \left(\prod_{k=3}^{n+2} a_k^2 \right) \left(1 + \sum_{k=3}^{n+2} \frac{1}{a_k^2} \right). \tag{8.10}$$

Finally, by (6.6), we get

$$d^2(f_0, V_n) = \frac{\det\big(\gamma(f_0, f_3, f_4, \ldots, f_{n+2})\big)}{\det\big(\gamma(f_3, f_4, \ldots, f_{n+2})\big)} = \left(1 + \sum_{k=3}^{n+2} \frac{1}{a_k^2}\right)^{-1} \to 0.$$

$\square$

8.2   The Height of an Infinite Parallelotope

Lemma 8.3. *Consider vectors $f_0, f_1, \ldots, f_m \in \mathbb{R}^\infty$ such that $f_k \notin l_2(\mathbb{N})$ for all $0 \leq k \leq m$. Denote by $f_r^{(n)} \in \mathbb{R}^n$ the projections of the vectors f_r on the subspace $\mathbb{R}^n$. Then, for all s, with $0 \leq s \leq m$,*

$$\frac{\Gamma(f_0, f_1, \ldots, f_m)}{\Gamma(f_0, \ldots, \hat{f}_s, \ldots, f_m)} := \lim_{n\to\infty} \frac{\Gamma(f_0^{(n)}, f_1^{(n)} \ldots, f_m^{(n)})}{\Gamma(f_0^{(n)}, \ldots, \widehat{f_s^{(n)}}, \ldots, f_m^{(n)})} = \infty, \quad (8.11)$$

if and only if for all $(C_k)_{k=0}^{m+1} \in \mathbb{R}^{m+1} \setminus \{0\}$ and $(C_0, \ldots, \widehat{C_s}, \ldots C_m) \in \mathbb{R}^m \setminus \{0\}$ holds

$$\sum_{r=0}^m C_r f_r \notin l_2(\mathbb{N}), \qquad \sum_{r=0, r\neq s}^m C_r f_r \notin l_2(\mathbb{N}). \quad (8.12)$$

Here, $\hat{f}_s$ means that the vector f_s is absent and $\Gamma(f_0, f_1, \ldots, f_m)$ is the Gram determinant.

Before proving Lemma 8.3, let us formulate one more statement.

Lemma 8.4. *Consider vectors $f_0, f_1, \ldots, f_m \in \mathbb{R}^\infty$ such that $\sum_{k=0}^m C_k f_k \notin l_2(\mathbb{N})$ for any non-trivial combination $(C_k)_{k=0}^m$. Then, for any s with $0 \leq s \leq m$, it holds that*

$$\frac{\det\big(I_{m+1} + \gamma(f_0, \ldots, f_m)\big)}{\det\big(I_m + \gamma(f_0, \ldots, \hat{f}_s, \ldots, f_m)\big)}$$

$$= \lim_{n\to\infty} \frac{\det\big(I_{m+1} + \gamma(f_0^{(n)}, \ldots, f_m^{(n)})\big)}{\det\big(I_m + \gamma(f_0^{(n)}, \ldots, \widehat{f_s^{(n)}}, \ldots, f_m^{(n)})\big)} = \infty. \quad (8.13)$$

Here, I_m is the identity matrix and $\gamma(f_0, \ldots, f_m)$ is the Gram matrix.

Proof. The proof follows from Lemma 8.3 and (3.8). $\qquad\square$

8.3 Particular Cases

8.3.1 Case: $m = 1$

Lemma 8.5 (Kosyak, 2018, 2019). *Consider vectors $f_0, f_1 \in \mathbb{R}^\infty$ such that $f_0, f_1 \notin l_2(\mathbb{N})$. Then,*

$$\frac{\Gamma(f_0, f_1)}{\Gamma(f_1)} = \lim_{n\to\infty} \frac{\Gamma(f_0^{(n)}, f_1^{(n)})}{\Gamma(f_1^{(n)})} = \infty, \quad \frac{\Gamma(f_0, f_1)}{\Gamma(f_0)} = \infty, \qquad (8.14)$$

if and only if $\quad C_0 f_0 + C_1 f_1 \notin l_2(\mathbb{N}) \quad$ *for all* $\quad (C_0, C_1) \in \mathbb{R}^2 \setminus \{0\}.$

$$(8.15)$$

Proof. The initial proof can be found in Kosyak (2018, 2019). Here, we give a different proof that *can be generalised for an arbitrary $m \in \mathbb{N}$.* For $t \in \mathbb{R}$ and $f_0, f_1 \in H$, where H is some Hilbert space, define the quadratic form

$$F_1(t) = \|t f_1 - f_0\|^2 = t^2 (f_1, f_1) - 2t(f_1, f_0) + (f_0, f_0)$$

$$= (A_1 t, t) - 2(t, b) + (f_0, f_0),$$

where $b = (f_1, f_0) \in \mathbb{R}$. We have

$$\min_{t\in\mathbb{R}} F_1(t) = F_1(t_0) = \frac{\Gamma(f_0, f_1)}{\Gamma(f_1)}, \quad \text{where} \quad t_0 = \frac{(f_1, f_0)}{(f_1, f_1)}, \quad (8.16)$$

and A_1 is a *Gram matrix* (see Definition 7.3):

$$A_1 = \gamma(f_1) = (f_1, f_1). \qquad (8.17)$$

Consider another Gram matrix:

$$B_1 := \gamma(f_0, f_1) = \begin{pmatrix} (f_0, f_0) & (f_0, f_1) \\ (f_1, f_0) & (f_1, f_1) \end{pmatrix}, \quad B_1^{(n)} := \gamma(f_0^{(n)}, f_1^{(n)}). \quad (8.18)$$

Then,

$$t_0 = \frac{(f_1, f_0)}{(f_1, f_1)} = \frac{-A_1^0(B_1)}{A_0^0(B_1)}, \qquad (8.19)$$

where $A_0^0(B_1)$ and $A_1^0(B_1)$ are the corresponding minors of the matrix B_1; see Definition 7.2. If we replace the vectors f_0, f_1 with $f_0^{(n)}, f_1^{(n)}$, the formulas (8.16) and (8.19) become

$$\min_{t \in \mathbb{R}} F_1^{(n)}(t) = F_1^{(n)}(t_0^{(n)}) = \frac{\Gamma(f_0^{(n)}, f_1^{(n)})}{\Gamma(f_1^{(n)})},$$

$$t_0^{(n)} = \frac{(f_1^{(n)}, f_0^{(n)})}{(f_1^{(n)}, f_1^{(n)})} = \frac{-A_1^0(B_1^{(n)})}{A_0^0(B_1^{(n)})}. \tag{8.20}$$

Suppose that there is an absolute constant C such that for all $n \in \mathbb{N}$,

$$\frac{\Gamma(f_0^{(n)}, f_1^{(n)})}{\Gamma(f_1^{(n)})} \leq C. \tag{8.21}$$

Without loss of generality, we can assume that for all $n \in \mathbb{N}$,

$$\Gamma(f_0^{(n)}) \leq C_1 \Gamma(f_1^{(n)}), \tag{8.22}$$

and we choose some subsequence n_k if necessary. Then, it is sufficient to verify only the first part of (8.14). Indeed, denote by $a_n = \Gamma(f_0^{(n)})$ and $b_n = \Gamma(f_1^{(n)})$. We have two sequences of positive numbers (a_n) and (b_n) with the property $\lim_n a_n = \lim_n b_n = \infty$. Let, for some absolute constant C_1, the inequality $a_n \leq C_1 b_n$ for all $n \in \mathbb{N}$ holds. Then, (8.22) holds. Suppose the opposite. Denote for all $N \in \mathbb{N}$, the set $E_N = \{n \in \mathbb{N} : a_n > N b_n\}$. Then, the set E_N is infinite for all N. Now, fix N, and denote all the elements of E_N by $(n_k)_{k=1}^{\infty}$. Then, $b_{n_k} < N^{-1} a_{n_k}$ for all $k \in \mathbb{N}$ or $\Gamma(f_1^{(n_k)}) \leq C_1 \Gamma(f_0^{(n_k)})$.

We prove that the sequence $t_0^{(n)}$, defined by (8.20), is bounded. Since all the matrices $\gamma(f_0^{(n)}, f_1^{(n)})$ are positively defined, we have

$$1 \geq \frac{(f_0^{(n)}, f_1^{(n)})^2}{(f_0^{(n)}, f_0^{(n)})(f_1^{(n)}, f_1^{(n)})} \overset{(8.22)}{\geq} \frac{(f_0^{(n)}, f_1^{(n)})^2}{C_1(f_1^{(n)}, f_1^{(n)})^2} = \frac{1}{C_1}(t_0^{(n)})^2.$$

Hence, the sequence $t_0^{(n)}$ is bounded. Therefore, there exists a subsequence $(t_0^{(n_k)})_{k \in \mathbb{N}}$ that converges to some $t \in \mathbb{R}$. This contradicts (8.21).

Indeed,

$$\lim_{n\to\infty} F_1^{(n)}(t) = \infty, \quad F_1^{(n)}(t_0^{(n)}) \le C, \quad \lim_{k\to\infty} t_0^{(n_k)} = t.$$

To prove the *necessity* condition, suppose that $C_0 f_0 + C_1 f_1 \in l_2(\mathbb{N})$ for some C_0, C_1. Let $f_1 = c_0 f_0 + h$, where $h \in l_2(\mathbb{N})$. We have

$$\Gamma(f_0, f_1) = \Gamma(f_0, h) \le \Gamma(f_0)\Gamma(h).$$

Since $h \in l_2(\mathbb{N})$ and $f_0 \notin l_2(\mathbb{N})$, we conclude that $\frac{\Gamma(f_0, f_1)}{\Gamma(f_0)}$ is bounded. $\square$

8.3.2 *Case: $m = 2$*

Lemma 8.6. *Let us consider three real vectors such that $f_0, f_1, f_2 \notin l_2(\mathbb{N})$. Then, for all r, s, with $0 \le r < s \le 2$, it holds that*

$$\frac{\Gamma(f_0, f_1, f_2)}{\Gamma(f_r, f_s)} := \lim_{n\to\infty} \frac{\Gamma(f_0^{(n)}, f_1^{(n)}, f_2^{(n)})}{\Gamma(f_r^{(n)}, f_s^{(n)})} = \infty, \tag{8.23}$$

if and only if $\sum_{r=0}^{2} C_r f_r \notin l_2(\mathbb{N})$ for all $(C_0, C_1, C_2) \in \mathbb{R}^3 \setminus \{0\}$ and $C_r f_r + C_s f_s \notin l_2(\mathbb{N})$ for all $(C_r, C_s) \in \mathbb{R}^2 \setminus \{0\}$.

Proof. Suppose that for all $n \in \mathbb{N}$,

$$\frac{\Gamma(f_0^{(n)}, f_1^{(n)}, f_2^{(n)})}{\Gamma(f_1^{(n)}, f_2^{(n)})} \le C. \tag{8.24}$$

Without loss of generality, we can assume that for all $n \in \mathbb{N}$,

$$\Gamma(f_0^{(n)}, f_1^{(n)}) \le C_2 \Gamma(f_1^{(n)}, f_2^{(n)}), \quad \Gamma(f_0^{(n)}, f_2^{(n)}) \le C_1 \Gamma(f_1^{(n)}, f_2^{(n)}), \tag{8.25}$$

and we choose some subsequence n_k, if necessary. Then, it is sufficient to verify (8.23) for $(r, s) = (1, 2)$.

For $t \in \mathbb{R}^2$ and $f_0, f_1, f_2 \in H$, where H is some Hilbert space, define the quadratic form $F_2(t)$ as follows:

$$F_2(t) = \left\| \sum_{r=1}^{2} t_r f_r - f_0 \right\|^2 = \sum_{k,r=1}^{2} t_k t_r (f_k, f_r) - 2 \sum_{k=1}^{2} t_k (f_k, f_0) + (f_0, f_0)$$

$$= (A_2 t, t) - 2(t, b) + (f_0, f_0),$$

where $b = (f_k, f_0)_{k=1}^{2} \in \mathbb{R}^2$ and A_2 is the *Gram matrix*:

$$A_2 = \gamma(f_1, f_2) = \left((f_k, f_r) \right)_{k,r=1}^{2}. \tag{8.26}$$

For t_0, defined by $A_2 t_0 = b$, we have by Lemma 6.3,

$$F_2(t) = (A_2 t, t) - 2(t, b) + (f_0, f_0) = (A_2(t - t_0), (t - t_0)) + \frac{\Gamma(f_0, f_1, f_2)}{\Gamma(f_1, f_2)};$$

therefore,

$$F_2(t_0) = \frac{\Gamma(f_0, f_1, f_2)}{\Gamma(f_1, f_2)}. \tag{8.27}$$

Consider the following matrix:

$$B_2 := \gamma(f_0, f_1, f_2) = \begin{pmatrix} (f_0, f_0) & (f_0, f_1) & (f_0, f_2) \\ (f_1, f_0) & (f_1, f_1) & (f_1, f_2) \\ (f_2, f_0) & (f_2, f_1) & (f_2, f_2) \end{pmatrix}. \tag{8.28}$$

By Cramer's rule (see Lemma 8.7), we have

$$t_0 = \frac{1}{A_0^0(B_2)} \begin{pmatrix} -A_1^0(B_2) \\ -A_2^0(B_2) \end{pmatrix}. \tag{8.29}$$

Let us consider $f_0, f_1, f_2 \in \mathbb{R}^\infty$, and let $f_0^{(n)}, f_1^{(n)}, f_2^{(n)}$ be their corresponding projections on the subspace $\mathbb{R}^n$. Define for $t \in \mathbb{R}^2$ the quadratic form $F_2^{(n)}(t)$ as follows:

$$F_2^{(n)}(t) = \left\| \sum_{r=1}^{2} t_r f_r^{(n)} - f_0^{(n)} \right\|^2 = (A_2^{(n)} t, t) - 2(t, b) + (f_0^{(n)}, f_0^{(n)}),$$

where $b = (f_k^{(n)}, f_0^{(n)})_{k=1}^2 \in \mathbb{R}^2$ and $A_2^{(n)}$ is the *Gram matrix*:

$$A_2^{(n)} = \gamma(f_1^{(n)}, f_2^{(n)}) = \left((f_k^{(n)}, f_r^{(n)}) \right)_{k,r=1}^2. \tag{8.30}$$

In what follows, we use a scalar product $(f, g)_H$ without referring to a specific H. By (8.45), we have

$$F_2^{(n)}(t_0^{(n)}) = \min_{t \in \mathbb{R}^2} F_2^{(n)}(t) = \frac{\Gamma(f_0^{(n)}, f_1^{(n)}, f_2^{(n)})}{\Gamma(f_1^{(n)}, f_2^{(n)})}.$$

We prove that the sequence $t_0^{(n)}$ is bounded. If we replace the vectors f_0, f_1, f_2 with $f_0^{(n)}, f_1^{(n)}, f_2^{(n)}$, we will get the following expressions:

$$t_0^{(n)} = \frac{1}{A_0^0(B_2^{(n)})} \begin{pmatrix} -A_1^0(B_2^{(n)}) \\ -A_2^0(B_2^{(n)}) \end{pmatrix}, \tag{8.31}$$

where $B^{(n)}(2)$ is defined by

$$B_2^{(n)} := \gamma(f_0^{(n)}, f_1^{(n)}, f_2^{(n)}) = \begin{pmatrix} (f_0^{(n)}, f_0^{(n)}) & (f_0^{(n)}, f_1^{(n)}) & (f_0^{(n)}, f_2^{(n)}) \\ (f_1^{(n)}, f_0^{(n)}) & (f_1^{(n)}, f_1^{(n)}) & (f_1^{(n)}, f_2^{(n)}) \\ (f_2^{(n)}, f_0^{(n)}) & (f_2^{(n)}, f_1^{(n)}) & (f_2^{(n)}, f_2^{(n)}) \end{pmatrix}. \tag{8.32}$$

Since all the matrices $B_2^{(n)}$ defined by (8.32) are positively defined, the inverse matrices $\left(B_2^{(n)} \right)^{-1}$ are also positively defined. We have the following expression for them, where we denote by B^T the *matrix transposed* to B:

$$\left(B_2^{(n)} \right)^{-1} = \frac{1}{\det B_2^{(n)}} \begin{pmatrix} A_0^0(B_2^{(n)}) & A_1^0(B_2^{(n)}) & A_2^0(B_2^{(n)}) \\ A_0^1(B_2^{(n)}) & A_1^1(B_2^{(n)}) & A_2^1(B_2^{(n)}) \\ A_0^2(B_2^{(n)}) & A_1^2(B_2^{(n)}) & A_2^2(B_2^{(n)}) \end{pmatrix}^T. \tag{8.33}$$

We prove that the sequence $t_0^{(n)}$, defined by (8.31),

$$t_0^{(n)} = \frac{1}{A_0^0(B_2^{(n)})} \begin{pmatrix} -A_1^0(B_2^{(n)}) \\ -A_2^0(B_2^{(n)}) \end{pmatrix}, \tag{8.34}$$

is bounded when (8.25) holds. Set

$$t_{rr}^{(n)} = A_r^r(B_2^{(n)}), \quad 0 \le r \le 2, \quad t_{rs}^{(n)} = -A_s^r(B_2^{(n)}), \quad 0 \le r \ne s \le 2. \tag{8.35}$$

Then, $t_0^{(n)} = \left(\frac{t_{01}^{(n)}}{t_{00}^{(n)}}, \frac{t_{02}^{(n)}}{t_{00}^{(n)}} \right)$. Since the matrix $\left(B_2^{(n)} \right)^{-1}$ is positively defined and (8.25) holds, we have

$$\left(t_{01}^{(n)} \right)^2 \le t_{00}^{(n)} t_{11}^{(n)}, \quad \left(t_{02}^{(n)} \right)^2 \le t_{00}^{(n)} t_{22}^{(n)}, \tag{8.36}$$

$$t_{22}^{(n)} \le C_2 t_{00}^{(n)}, \quad t_{11}^{(n)} \le C_1 t_{00}^{(n)}, \tag{8.37}$$

$$\|t_0^{(n)}\|^2 := \frac{|t_{01}^{(n)}|^2 + |t_{02}^{(n)}|^2}{|t_{00}^{(n)}|^2} \overset{(8.36)}{\le} \frac{t_{11}^{(n)} + t_{22}^{(n)}}{t_{00}^{(n)}} \tag{8.38}$$

$$\overset{(8.37)}{\le} \frac{C_1 t_{00}^{(n)} + C_2 t_{00}^{(n)}}{t_{00}^{(n)}} = C_1 + C_2. \tag{8.39}$$

Hence, the sequence $t_0^{(n)} \in \mathbb{R}^2$ is bounded. Therefore, there exists a subsequence $(t_0^{(n_k)})_{k \in \mathbb{N}}$ that converges to some $t \in \mathbb{R}^2$. This contradicts (8.24). Indeed,

$$\lim_{n \to \infty} F_2^{(n)}(t) = \infty, \quad F_2^{(n)}(t_0^{(n)}) \le C, \quad \lim_{k \to \infty} t_0^{(n_k)} = t.$$

We prove the *necessity* condition for $(r, s) = (0, 1)$, with the general case being similar. Suppose that $\sum_{k=0}^{2} C_k f_k \in l_2(\mathbb{N})$ for some $(C_k)_k$ but $C_0 f_0 + C_1 f_1 \notin l_2(\mathbb{N})$ for all $(C_0, C_1) \in \mathbb{R}^2 \setminus \{0\}$. Let $f_2 = c_0 f_0 + c_1 f_1 + h$, where $h \in l_2(\mathbb{N})$. We have

$$\Gamma(f_0, f_1, f_2) = \Gamma(f_0, f_1, h) \le \Gamma(f_0, f_1)\Gamma(h).$$

Since $h \in l_2(\mathbb{N})$ but $C_0 f_0 + C_1 f_1 \notin l_2(\mathbb{N})$ for all $(C_0, C_1) \in \mathbb{R}^2 \setminus \{0\}$, we conclude that $\frac{\Gamma(f_0, f_1, f_2)}{\Gamma(f_0, f_1)}$ is bounded. $\qquad\square$

8.4 Proof of Lemma 8.3

Proof. Suppose that for all $n \in \mathbb{N}$, we have

$$\frac{\Gamma(f_0^{(n)}, f_1^{(n)} \ldots, f_m^{(n)})}{\Gamma(f_1^{(n)}, f_2^{(n)} \ldots, f_m^{(n)})} \leq C. \tag{8.40}$$

Without loss of generality, we can assume that for all $n \in \mathbb{N}$ and $1 \leq s \leq m$, it holds that

$$\Gamma(f_0^{(n)}, \ldots, \widehat{f_s^{(n)}}, \ldots, f_m^{(n)}) \leq C_s \Gamma(f_1^{(n)}, f_2^{(n)}, \ldots, f_m^{(n)}), \tag{8.41}$$

and we choose some subsequence n_k if necessary. Consider the following quadratic forms of $t = (t_r)_{r=1}^m \in \mathbb{R}^m$:

$$F_m^{(n)}(t) = \|f_0^{(n)} - \sum_{r=1}^m t_r f_r^{(n)}\|^2. \tag{8.42}$$

The forms $F_m^{(n)}(t)$ defined by (8.42) have the following properties. For any fixed $t \in \mathbb{R}^m$, we have $\lim_{n \to \infty} F_m^{(n)}(t) = \infty$. By Lemma 6.3, there exists some $t_0^{(n)} \in \mathbb{R}^m$ such that

$$F_m^{(n)}(t_0^{(n)}) = \min_{t \in \mathbb{R}^m} F_m^{(n)}(t) = \frac{\Gamma(f_0^{(n)}, f_1^{(n)}, \ldots, f_m^{(n)})}{\Gamma(f_1^{(n)}, f_2^{(n)} \ldots, f_m^{(n)})}. \tag{8.43}$$

We prove that the sequence $t_0^{(n)}$ is bounded. To find $t_0^{(n)}$ explicitly in (8.43), we introduce some notations. For $t \in \mathbb{R}^m$ and $f_0, f_1, \ldots, f_m \in H$, define the function

$$F_m(t) = \left\| \sum_{k=1}^m t_k f_k - f_0 \right\|^2 = \sum_{k,r=1}^m t_k t_r (f_k, f_r)_H$$

$$- 2 \sum_{k=1}^m t_k (f_k, f_0)_H + (f_0, f_0)_H$$

$$= (A_m t, t)_{\mathbb{R}^m} + 2(t, b)_{\mathbb{R}^m} + (f_0, f_0)_H,$$

where $b = \left(\left((f_k, f_0)_H\right)\right)_{k=1}^m \in \mathbb{R}^m$ and A_m is the *Gram matrix*, referring to Definition 7.3:

$$A_m = \gamma(f_1, \ldots, f_m) = \left((f_k, f_r)\right)_{k,r=1}^m. \tag{8.44}$$

The minimum of $F_m(t)$ is attained at t_0, defined by $A_m t_0 = b$. By Remark 6.1, (6.11) and (6.8), we get

$$F_m(t) = (A_m t, t) - 2(t, b) + (f_0, f_0)$$

$$= (A_m(t - t_0), (t - t_0)) + \frac{\Gamma(f_0, f_1, \ldots, f_m)}{\Gamma(f_1, \ldots, f_m)},$$

$$\text{and therefore,} \quad F_m(t_0) = \frac{\Gamma(f_0, f_1, \ldots, f_m)}{\Gamma(f_1, \ldots, f_m)}. \tag{8.45}$$

Consider the following matrix:

$$B_m = \gamma(f_0, f_1, \ldots, f_m) = \begin{pmatrix} (f_0, f_0) & (f_0, f_1) & (f_0, f_2) & \cdots & (f_0, f_m) \\ (f_1, f_0) & (f_1, f_1) & (f_1, f_2) & \cdots & (f_1, f_m) \\ (f_2, f_0) & (f_2, f_1) & (f_2, f_2) & \cdots & (f_2, f_m) \\ & & & \vdots & \\ (f_m, f_0) & (f_m, f_1) & (f_m, f_2) & \cdots & (f_m, f_m) \end{pmatrix}.$$

By Cramer's rule (see Lemma 8.7), the solution to $A_m t_0 = b$ is as follows:

$$t_0 = \frac{1}{A_0^0(B_m)} \begin{pmatrix} -A_1^0(B_m) \\ -A_2^0(B_m) \\ \vdots \\ -A_m^0(B_m) \end{pmatrix}. \tag{8.46}$$

If we replace the vectors $(f_k)_{k=0}^m$ with $\left(f_k^{(n)}\right)_{k=0}^m$, we will get the following expression:

$$t_0^{(n)} = \frac{1}{A_0^0(B_m^{(n)})} \begin{pmatrix} -A_1^0(B_m^{(n)}) \\ \vdots \\ -A_m^0(B_m^{(n)}) \end{pmatrix}, \tag{8.47}$$

where $B^{(n)}(m)$ is defined by

$$B_m^{(n)} = \gamma(f_0^{(n)}, \ldots, f_m^{(n)}) = \begin{pmatrix} (f_0^{(n)}, f_0^{(n)}) & (f_0^{(n)}, f_1^{(n)}) & \cdots & (f_0^{(n)}, f_m^{(n)}) \\ (f_1^{(n)}, f_0^{(n)}) & (f_1^{(n)}, f_1^{(n)}) & \cdots & (f_1^{(n)}, f_m^{(n)}) \\ & & \vdots & \\ (f_m^{(n)}, f_0^{(n)}) & (f_m^{(n)}, f_1^{(n)}) & \cdots & (f_m^{(n)}, f_m^{(n)}) \end{pmatrix}.$$

Since all the matrices $B_m^{(n)}$ are positively defined, the inverse matrices $\left(B_m^{(n)}\right)^{-1}$ are also positively defined. We have the following expression for them:

$$\left(B_m^{(n)}\right)^{-1} = \frac{1}{\det B_m^{(n)}} \begin{pmatrix} A_0^0(B_m^{(n)}) & A_1^0(B_m^{(n)}) & \cdots & A_m^0(B_m^{(n)}) \\ A_0^1(B_m^{(n)}) & A_1^1(B_m^{(n)}) & \cdots & A_m^1(B_m^{(n)}) \\ & & \vdots & \\ A_0^m(B_m^{(n)}) & A_1^m(B_m^{(n)}) & \cdots & A_m^m(B_m^{(n)}) \end{pmatrix}^T.$$

$$(8.48)$$

We prove that the sequence $\left(t_0^{(n)}\right)_n$ is bounded when (8.41) holds:

$$t_0^{(n)} = \frac{1}{A_0^0(B_m^{(n)})} \begin{pmatrix} -A_1^0(B_m^{(n)}) \\ \vdots \\ -A_m^0(B_m^{(n)}) \end{pmatrix}. \tag{8.49}$$

Set $t_{rr}^{(n)} = A_r^r(B_m^{(n)})$, $0 \leq r \leq m$, $t_{rs}^{(n)} = -A_s^r(B_m^{(n)})$, $0 \leq r \neq s \leq m$.

$$(8.50)$$

Then, $t_0^{(n)} = \left(\frac{t_{01}^{(n)}}{t_{00}^{(n)}}, \frac{t_{02}^{(n)}}{t_{00}^{(n)}}, \ldots \frac{t_{0m}^{(n)}}{t_{00}^{(n)}} \right)$. Since the matrix $\left(B_m^{(n)}\right)^{-1}$ is positively defined and (8.25) holds, we have

$$\left(t_{rs}^{(n)}\right)^2 \leq t_{rr}^{(n)} t_{ss}^{(n)}, \quad \text{for all} \quad 0 \leq r < s \leq m, \tag{8.51}$$

$$t_{ss}^{(n)} \leq C_s t_{00}^{(n)}, \quad \text{for all} \quad 0 \leq s \leq m-1, \tag{8.52}$$

$$\|t_0^{(n)}\|^2 := \frac{\sum_{s=1}^m |t_{0s}^{(n)}|^2}{|t_{00}^{(n)}|^2} \overset{(8.51)}{\le} \frac{\sum_{s=1}^m t_{ss}^{(n)}}{t_{00}^{(n)}} \tag{8.53}$$

$$\overset{(8.52)}{\le} \frac{\sum_{s=1}^m C_s t_{00}^{(n)}}{t_{00}^{(n)}} = \sum_{s=1}^m C_s. \tag{8.54}$$

Hence, the sequence $t_0^{(n)} \in \mathbb{R}^m$ is bounded. Therefore, there exists a subsequence $(t_0^{(n_k)})_{k \in \mathbb{N}}$ that converges to some $t \in \mathbb{R}^m$. This contradicts (8.40). Indeed,

$$\lim_{n \to \infty} F_m^{(n)}(t) = \infty, \quad F_m^{(n)}(t_0^{(n)}) \le C, \quad \lim_{k \to \infty} t_0^{(n_k)} = t.$$

The necessity is proved as in the cases of $m = 1$ and $m = 2$. $\square$

8.5   Reformulation of Cramer's Rule

Consider two *Gram matrices*, referring to Definition 7.3:

$$A_m = \gamma(f_1, \ldots, f_m) = \left((f_k, f_r)\right)_{k,r=1}^m, \quad B_m = \gamma(f_0, f_1, \ldots, f_m), \tag{8.55}$$

and a vector $b = (f_k, f_0)_{k=1}^m \in \mathbb{R}^m$. The solution to the equation $A_m t = b$ is as follows.

Lemma 8.7. *We have*

$$t = A_m^{-1} b = \frac{1}{A_0^0(B_m)} \begin{pmatrix} -A_1^0(B_m) \\ -A_2^0(B_m) \\ \vdots \\ -A_m^0(B_m) \end{pmatrix}. \tag{8.56}$$

Proof.   By *Cramer's rule*, the solutions for a system of linear equations

$$At = b, \tag{8.57}$$

where $A \in \mathrm{Mat}(m, \mathbb{C})$, with $\det A \ne 0$ and $t, b \in \mathbb{C}^m$, are given by the following formulas:

$$t_k = \frac{\det(A_k)}{\det(A)}, \quad 1 \le k \le m, \tag{8.58}$$

where A_k is the matrix formed by replacing the kth column of $A := A_m$ by the column vector b. Consider the matrix B_m defined by (8.55). We have

$$\det (A) = A_0^0(B_m), \quad \det (A_k) = -A_k^0(B_m), \quad 1 \le k \le m, \qquad (8.59)$$

thus implying (8.56). Recall that $A_s^r(C)$ denote the *cofactors* of the matrix C; see Definition 7.4. $\qquad\square$

Chapter 9

Representations of the Group $\mathrm{GL}_0(2\infty, \mathbb{R})$: Studying the Case of $m > 3$

To prove the irreducibility of the representation $T^{R,\mu,m}$, defined by (3.6) for general $m \in \mathbb{N}$, we need:

(1) to know the minimal generating set of conditions for the orthogonality $(\mu^m_{(b,a)})^{L_t} \perp \mu^m_{(b,a)}$ for all $t \in \mathrm{GL}(m, \mathbb{R}) \setminus \{e\}$; see Chapter 3.3. In fact, it is sufficient to replace the group $\mathrm{GL}(m, \mathbb{R})$ by $\pm\mathrm{SL}(m, \mathbb{R})$; see Remark 3.9. These conditions will be expressed in terms of some divergent series, $\big(S_\beta(\mu)\big)_{\beta \in B}$.

(2) to find an appropriate combination of generators A_{kn} of all one-parameter subgroups or an appropriate function of generators. This combination will be expressed in terms of a divergent series, $\big(\Sigma_\alpha(A)\big)_{\alpha \in A}$.

(3) to show that (1) implies (2).

What we know now are the following:

(1) If we have some continuous finite-dimensional group G acting on an infinite-dimensional space X with a measure μ and we are interested in the *"admissible"* action, i.e., $\mu^{\alpha_t} \sim \mu$ for every $t \in G$, the problem is much easier, where $\alpha : G \to \mathrm{Aut}(X)$. To find a minimal set, it is sufficient to verify the equivalence only on the one-parameter

143

subgroups $g_k(t)$, generating the group G. This follows from the *transitivity* of the relation of the equivalence on the sets of probability measures $\mu \sim \nu$. But the relation of the orthogonality on the sets of measures is not *transitive*. That is why the description of the minimal set is so complicated. When $m = 1$, the minimal subset is reduced to $-1 \in \mathrm{GL}(1, \mathbb{R})$.

When $m = 2$, the description of the minimal generating set is given in Remark 3.7. The families (3.21) are one-parameter subgroups, the families (3.22) are simply reflections of (3.21) and the family (3.23) depends on two parameters. All elements are of order 2 except the elements in subgroups given in (3.21).

When $m = 3$, the description of the minimal generating set is given in Lemma 3.7, and it involves the families (3.38)–(3.50) depending respectively on two, three and five parameters; see the remarks after Lemma 3.6.

When $m = 4$, we still do not know the answer.

(2) When $m = 1$ and $m = 2$, it was sufficient for the approximation of x_{kr} or D_{kr} to use linear combinations of products of two generators, $A_{kn}A_{rn}$, for $n \in \mathbb{Z}$. When $m = 3$, we were not able to use only quadratic functions of generators. As Lemma 5.4 demonstrates, we were forced to use $\exp\big(is_k(x_{rk} - a_{rk})\big)A_{kn}$ in order to approximate D_{rn}.

(3) This relies mainly on the properties of the generalised characteristic polynomial, explicit expression for the minimum of the quadratic form restricted to a hyperplane and a theorem regarding the height of an infinite parallelotope. The last two points hold for general m. But the approximation of x_{rk} and D_{rk} should be done in addition.

Chapter 10

Another Interesting Example

10.1 Group $B_0^{\mathbb{N}}$, Arbitrary Measure μ

Let $B_0^{\mathbb{N}}$ be the group of finite real upper-triangular matrices with unities on the principal diagonal, and let $B^{\mathbb{N}}$ be the group of all such matrices (not necessarily finite):

$$B_0^{\mathbb{N}} = \left\{ I + x = I + \sum_{k<n} x_{kn} E_{kn} \mid x \text{ is finite} \right\},$$

$$B^{\mathbb{N}} = \left\{ I + x = I + \sum_{k<n} x_{kn} E_{kn} \mid x \text{ is arbitrary} \right\}.$$

Let μ be an arbitrary probability measure on the group $B^{\mathbb{N}}$. If $\mu^{R_t} \sim \mu$ and $\mu^{L_t} \sim \mu$ for all $t \in B_0^{\mathbb{N}}$, an analogue of the right $T^{R,\mu}$ and left $T^{L,\mu}$ regular representations of the group $B_0^{\mathbb{N}}$, i.e., $T^{R,\mu}$, $T^{L,\mu} : B_0^{\mathbb{N}} \to U(H_\mu)$, are defined in the space $H_\mu = L^2(B^{\mathbb{N}}, \mu)$ by (2.2) and (2.3), respectively. For the generators $A_{kn}^{R,\mu}$ $(A_{kn}^{L,\mu})$ of the one-parameter groups $I + t E_{kn}$, $t \in \mathbb{R}$, $k < n$, corresponding to the right $T^{R,\mu}$ (resp., the left $T^{L,\mu}$) regular

145

representation, we have the following formulas:

$$A_{kn}^{R,\mu} = \frac{d}{dt} T_{I+tE_{kn}}^{R,\mu} \Big|_{t=0} = \sum_{r=1}^{k-1} x_{rk} D_{rn}(\mu) + D_{kn}(\mu), \qquad (10.1)$$

$$A_{kn}^{L,\mu} = \frac{d}{dt} T_{I+tE_{kn}}^{L,\mu} \Big|_{t=0} = -\left(D_{kn}(\mu) + \sum_{m=n+1}^{\infty} x_{nm} D_{km}(\mu) \right), \qquad (10.2)$$

where $D_{kn}(\mu) = \frac{\partial}{\partial x_{kn}} + \frac{d}{dt}\left(\frac{d\mu(x(I+tE_{kn}))}{d\mu(x)} \right)^{1/2} \Big|_{t=0}$. For an arbitrary product measure $\mu = \bigotimes_{k<n} \mu_{kn}$, we have

$$D_{kn}(\mu) = \frac{\partial}{\partial x_{kn}} + \frac{\partial}{\partial x_{kn}}\left(\ln \mu_{kn}^{1/2}(x_{kn}) \right), \qquad (10.3)$$

where we write $d\mu_{kn}(x) = \mu_{kn}(x)dx$, $x \in \mathbb{R}$.

10.2 Group $B_0^{\mathbb{N}}$, Gaussian-Centred Measure

Let us define the Gaussian product measure μ_b on the group $B^{\mathbb{N}}$ in the following way (see details in Kosyak, 2018, Chapter 2.1):

$$d\mu_b(x) = \bigotimes_{k<n}(b_{kn}/\pi)^{1/2} \exp(-b_{kn}x_{kn}^2)dx_{kn} = \bigotimes_{k<n} d\mu_{b_{kn}}(x_{kn}), \qquad (10.4)$$

where $b = (b_{kn})_{k<n}$ is some set of positive numbers. In this case, we have

$$A_{kn}^{R,\mu} = \frac{d}{dt} T_{I+tE_{kn}}^{R,\mu} \Big|_{t=0} = \sum_{r=1}^{k-1} x_{rk} D_{rn} + D_{kn}, \quad D_{kn} = \frac{\partial}{\partial x_{kn}} - b_{kn}x_{kn}. \qquad (10.5)$$

It turns out that the measure μ_b is always $B_0^{\mathbb{N}}$-right quasi-invariant. Therefore, we can construct a family of analogues of the right T^{R,μ_b} and left T^{L,μ_b} (if the measure μ_b is $B_0^{\mathbb{N}}$-left quasi-invariant) regular representations of the group $B_0^{\mathbb{N}}$ in the space $L_2(B^{\mathbb{N}}, \mu_b)$. They are defined by (2.2) and (2.3), respectively.

Theorem 10.1 (Kosyak, 1992, 2018). *The right regular representation T^{R,μ_b} of the group $B_0^{\mathbb{N}}$ is irreducible if and only if:*

(1) $\mu^{L_t} \perp \mu$ *for all* $t \in B_0^{\mathbb{N}} \setminus \{e\}$;
(2) *the measure μ is $B_0^{\mathbb{N}}$-ergodic.*

Lemma 10.1 (Kosyak, 2018, Lemma 2.1.6). *We have $\mu_b^{L_t} \perp \mu_b$ for all* $t \in B_0^{\mathbb{N}} \setminus \{e\}$, *if and only if*

$$S_{kn}^L(\mu_b) = \sum_{m=k+1}^{\infty} \frac{b_{km}}{b_{nm}} = \infty \quad \text{for all} \quad k < n. \tag{10.6}$$

Idea of the proof of irreducibility; for details, see Kosyak (1992, 2018).

Conditions (1) and (2) are necessary conditions for the irreducibility of the representation T^{R,μ_b}. We show that they are also sufficient. Let $\mathfrak{A}(B_0^{\mathbb{N}})$ be a von Neumann algebra generated by the representation T^{R,μ_b}:

$$\mathfrak{A}(B_0^{\mathbb{N}}) = \left(T_t^{R,\mu_b} \mid t \in B_0^{\mathbb{N}} \right)''. \tag{10.7}$$

To prove irreducibility, it is sufficient to show that $U_{kn}(t) \in \mathfrak{A}(B_0^{\mathbb{N}})$ for all $k, n \in \mathbb{N}$, $k < n$, where $U_{kn}(t) = e^{itx_{kn}}$. In this case, we have

$$L^{\infty}(B^{\mathbb{N}}, \mu_b) \subset \mathfrak{A}(B_0^{\mathbb{N}}),$$

$$\text{hence} \quad \left(\mathfrak{A}(B_0^{\mathbb{N}}) \right)' \subset \left(L^{\infty}(B^{\mathbb{N}}, \mu_b) \right)' = L^{\infty}(B^{\mathbb{N}}, \mu_b), \tag{10.8}$$

since the algebra $L^{\infty}(B^{\mathbb{N}}, \mu_b)$ is *maximal abelian*. Suppose now that some bounded operator A commutes with the representation $[T_t^{R,\mu_b}, A] = 0$ for all $t \in B_0^{\mathbb{N}}$. Then, by (10.8), $A \in L^{\infty}(B^{\mathbb{N}}, \mu_b)$, i.e., A is a multiplication operator by some function $a \in L^{\infty}(B^{\mathbb{N}}, \mu_b)$. The commutation $[T_t^{R,\mu_b}, a] = 0$ implies $a(xt) = a(x)$ a.e. $\bmod \mu_b$. By the ergodicity of the measure μ_b on $B^{\mathbb{N}}$, we conclude that $a(x) = const$, and hence $A = CI$, i.e., the representation T^{R,μ_b} is irreducible.

To illustrate the approximation, we show here only that $e^{itx_{12}} \in \mathfrak{A}(B_0^{\mathbb{N}})$, or $x_{12} \, \eta \, \mathfrak{A}(B_0^{\mathbb{N}})$, i.e., the operator x_{12} is *affiliated* with the algebra $\mathfrak{A}(B_0^{\mathbb{N}})$. Recall (see Dixmier, 1969) that, a not-necessarily-bounded self-adjoint operator A in a Hilbert space H is said to be *affiliated* with a von Neumann algebra M of operators in this Hilbert space H if $e^{itA} \in M$ for all $t \in \mathbb{R}$. We write $A \, \eta \, M$. We show that the operator x_{12} can be approximated in the *strong resolvent sense* by a linear combination of the following operators $A_{1k}A_{2k}$, $k \geq 3$. Using (10.5), we get

$$A_{1k}A_{2k} = D_{1n}(x_{12}D_{1k} + D_{2k}) = x_{12}D_{1k}^2 + D_{1k}D_{2k} \quad k \geq 3. \qquad (10.9)$$

Lemma 10.2 (Kosyak, 2018). *The convergence* $\sum_{k=N_1}^{N_2} t_k A_{1k}A_{2k} \to x_{12}$ *holds if and only if* $S_{12}^L(\mu_b) = \sum_{k=3}^{\infty} \frac{b_{1k}}{b_{2k}} = \infty$.

Proof. Since $MD_{1k}^2 1 = \frac{b_{1k}}{2}$, we should choose $\sum_{k=3}^{n+2} t_k \frac{b_{1k}}{2} = 1$. In this case, we have

$$\left\| \sum_{k=3}^{n+2} t_k (A_{1k}A_{2k} - x_{12})1 \right\|^2 = \left\| x_{12} \sum_{k=3}^{n+2} t_k \left[\left(D_{1k}^2 - \frac{b_{1k}}{2} \right) + D_{1k}D_{2k} \right] 1 \right\|^2$$

$$= \left\| x_{12} \sum_{k=3}^{n+2} t_k f_k \right\|^2 = \|x_{12}\|^2 \left\| \sum_{k=3}^{n+2} t_k f_k \right\|^2$$

$$= \|x_{12}\|^2 \sum_{k=3}^{n+2} t_k^2 \|f_k\|^2$$

$$= \|x_{12}\|^2 \times \sum_{k=3}^{n+2} t_k^2 \left(\frac{b_{1k}^2}{2} + \frac{b_{1k}b_{2k}}{4} \right), \qquad (10.10)$$

$$\text{where} \quad f_k = \left[\left(D_{1k}^2 - \frac{b_{1k}}{2} \right) + D_{1k}D_{2k} \right] 1.$$

Therefore, we get

$$\|f_k\|^2 = \left(\frac{b_{1k}^2}{2} + \frac{b_{1k}b_{2k}}{4} \right) \quad \text{and} \quad f_k \perp f_s \text{ for } k \neq s.$$

By (10.10) and (6.2), we get

$$\min_{t \in \mathbb{R}^n} \left(\sum_{k=3}^{n+2} \|f_k\|^2 t_k^2 \mid \sum_{k=3}^{n+2} \frac{t_k b_{1k}}{2} = 1 \right) = \left(\sum_{k=3}^{n+2} \frac{b_{1k}^2}{4\|f_k\|^2} \right)^{-1} \to 0$$

$$\text{when} \quad n \to \infty,$$

since

$$\sum_{k=3}^{n+2}\frac{b_{1k}^2}{4\|f_k\|^2} = \sum_{k=3}^{n+2}\frac{b_k^2}{4\left(\frac{b_{1k}^2}{2}+\frac{b_{1k}b_{2k}}{4}\right)} \sim \sum_{k=3}^{n+2}\frac{b_{1k}}{b_{2k}} \to \infty, \quad \text{by} \quad S_{12}^L(\mu_b)=\infty.$$

This is precisely the condition of orthogonality $\mu_b^{L_t} \perp \mu_b$; see Lemma 10.1. $\qquad\square$

We give here a more conceptual proof of this fact. Using the appropriate *Fourier transform* F_2 in the variables $(x_{1k}, x_{2k})_{k=3}^\infty$, referring to details in Kosyak (2018, Section 2.1.3, formula (2.15)), we get $F_2(D_{1k}) = y_{1k}$, $F_2(D_{2k}) = y_{2k}$, $k \geq 3$; therefore,

$$F_2(A_{1k}A_{2k}) = x_{12}y_{1k}^2 + y_{1k}y_{2k}. \tag{10.11}$$

The corresponding measure $\mu_{1/4b}(y)$ in the variables $(y_{1k}, y_{2k})_{k=3}^\infty$ is defined by

$$d\mu_{1/4b}(y) = \bigotimes_{k=1}^2\bigotimes_{n=3}^\infty \sqrt{\frac{1}{4b_{kn}\pi}} \exp\left(-\frac{y_{kn}^2}{4b_{kn}}\right) dy_{kn} = \bigotimes_{k=1}^2\bigotimes_{n=3}^\infty d\mu_{1/4b_{kn}}(y_{kn}). \tag{10.12}$$

The corresponding canonical measure $\mu_{1/2}(z)$ is as follows:

$$d\mu_{1/2}(z) = \bigotimes_{k=1}^2\bigotimes_{n=3}^\infty \sqrt{\frac{1}{2\pi}} \exp\left(-\frac{z_{kn}^2}{2}\right) dz_{kn} = \bigotimes_{k=1}^2\bigotimes_{n=3}^\infty d\mu_{1/2}(z_{kn}). \tag{10.13}$$

In the *canonical coordinates* z_{kn}, the expression $F_2(A_{1k}A_{2k})$ will have the following form:

$$x_{12}2b_{1k}z_{1k}^2 + 2\sqrt{b_{1k}b_{2k}}z_{1k}z_{2k} = 2\sqrt{b_{1k}}\left(x_{12}z_{1k}^2 + a_k z_{1k}z_{2k}\right),$$
$$a_k = \sqrt{b_{2k}/b_{1k}}.$$

Let us denote by $\langle f_n \mid n \in \mathbb{N}\rangle$ the *closure of the linear space* generated by the set of vectors $(f_n)_{n\in\mathbb{N}}$ in a Hilbert space H.

Lemma 10.3. *Set $f_0 = x_{12}, f_k = x_{12}z_{1k}^2 + a_k z_{1k} z_{2k}$. We have $f_0 \in \langle f_k \mid k \geq 3 \rangle$ if and only if $\sum_{k=3}^{\infty} \frac{1}{a_k^2} = \infty$.*

Proof. We repeat here the proof of Lemma 8.2. Consider the hyperplane V_n generated by n vectors $f_3, \ldots, f_{n+2}$. By Lemma 6.3, we have

$$d^2(f_0, V_n) = \frac{\Gamma(f_0, f_3, f_4, \ldots, f_{n+2})}{\Gamma(f_3, f_4, \ldots, f_{n+2})}. \tag{10.14}$$

Further,

$$\gamma(f_0, f_3, f_4, \ldots, f_{n+2}) = \begin{pmatrix} 1 & 1 & \cdots & 1 \\ 1 & 1 + a_3^2 & \cdots & 1 \\ & & \vdots & \\ 1 & 1 & \cdots & 1 + a_{n+2}^2 \end{pmatrix} \tag{10.15}$$

and

$$\gamma(f_3, f_4, \ldots, f_{n+2}) = \begin{pmatrix} 1 + a_3^2 & 1 & \cdots & 1 \\ 1 & 1 + a_4^2 & \cdots & 1 \\ & & \vdots & \\ 1 & 1 & \cdots & 1 + a_{n+2}^2 \end{pmatrix}. \tag{10.16}$$

Finally, by (7.29), we get

$$d^2(f_0, V_n) = \frac{\det\big(\gamma(f_0, f_3, f_4, \ldots, f_{n+2})\big)}{\det\big(\gamma(f_3, f_4, \ldots, f_{n+2})\big)}$$

$$\overset{(7.29)}{=} \frac{\left(\prod_{k=3}^{n+2} a_k^2\right)}{\left(\prod_{k=3}^{n+2} a_k^2\right)\left(1 + \sum_{k=3}^{n+2} \frac{1}{a_k^2}\right)} = \left(1 + \sum_{k=3}^{n+2} \frac{1}{a_k^2}\right)^{-1}. \qquad \square$$

Bibliography

Akhiezer, N. I. and Glazman, I. M. (1993). *Theory of Linear Operators in Hilbert Space*, Dover Books on Mathematics, (Original Russian edition, Moscow: Nauka, 1966).

Albeverio, S. and Kosyak, A. (2006). Quasiregular representations of the infinite-dimensional nilpotent group, *J. Funct. Anal.* **236**, pp. 634–681.

Baik, J., Deift, P. and Johansson, K. (1999). On the distribution of the length of the longest increasing subsequence of random permutations, *J. Amer. Math. Soc.* **12**, pp. 1119–1178.

Beckenbach, E. F. and Bellman, R. (1961). *Inequalities*, Springer, Berlin, Göttingen, Heidelberg.

Berezanskii, Yu. M. (1986). *Selfadjoint Operators in Spaces of Functions of Infinitely Many Variables*, Translated from the Russian by H. H. McFadden, Translations of Mathematical Monographs, Vol. 63, American Mathematical Society, Providence, RI.

Borodin, A., Okounkov, A. and Ol'shanskiĭ, G. (2000). Asymptotics of Plancherel measures for symmetric groups, *J. Amer. Math. Soc.* **13**, pp. 481–515 (electronic).

Borodin, A. and Ol'shanskiĭ, G. (1998). Point processes and the infinite symmetric group, *Math. Res. Lett.* **5**, pp. 799–816.

Bufetov, A. I. (2014). Finiteness of ergodic unitarily invariant measures on spaces of infinite matrices. *Annales de l'Institut Fourier.* **64**(3), 893–907.

Dixmier, J. (1969). *Les algèbres d'opérateurs dans l'espace hilbertien*, 2nd edition, Gauthier-Villars, Paris.

Dixmier, J. (1969). *Les C*-algèbres et leur représentations,* Gautier-Villars, Paris.

Engel, G. M. and Schneider, H. (1976). The Hadamard-Fischer inequality for a class of matrices defined by eigenvalue monotonicity, *Linear Multilinear Algebra* **4**, pp. 155–176.

Gantmacher, R. F. (1958). *Matrizenrechnung,* Teil 1, Veb Deutscher Verlag der Wissenschaften, Berlin.

Gram, J. P. (1879). *On Raekkeudviklinger bestemmte ved Hjaelp of de mindste Kvadraters Methode,* Copenhagen.

Haar, A. (1933). Der Massbegriff in der Theorie der kontinuierlichen Gruppen, *Ann. Math.* **34**(1), pp. 147–169.

Horn, R. A. and Johnson, C. R. (1989). *Matrix Analysis,* Cambridge University Press, Cambridge.

Horn, R. A. and Johnson, C. R. (1989). *Topics in Matrix Analysis,* Cambridge University Press, Cambridge.

Hou, S. H. (1998). Classroom Note: A Simple Proof of the Leverrier–Faddeev Characteristic Polynomial Algorithm, *SIAM Rev.* **40**(3), pp. 706–709.

Ismagilov, R. S. (1996). *Representations of Infinite-Dimensional Groups,* in: Translations of Mathematical Monographs, Vol. **152**, American Mathematical Society, Providence, RI.

Kakutani, S. (1948). On equivalence of infinite product measures, *Ann. Math.* **4**, pp. 214–224.

Kerov, S., Ol'shanskiĭ, G. and Vershik, A. (1993). Harmonic analysis on the infinite symmetric group. A deformation of the regular representation, *C. R. Acad. Sci. Paris Sér. 1.* **316**, pp. 773–778.

Khrennikov, Y., Kosyak, A. V. and Shelkovich, V. M. (2012). Wavelet analysis on adeles and pseudo-differential operators, *J. Fourier. Anal. Appl.* **18**(6), pp. 1215–1264.

Kirillov, A. A. (1962). Unitary representations of nilpotent Lie groups, *Usp. Mat. Nauk.* **17**(4), pp. 57–110.

Kirillov, A. A. (1973). Representations of infinite-dimensional unitary groups, *Dokl. Akad. Nauk SSSR.* **212**(2), pp. 288–290, English Transl.: *Sov. Math. Dokl.* **14**, pp. 1355–1358 (1974).

Kosyak, A. V. (1990). Irreducibility criterion for regular Gaussian representations of groups of finite upper triangular matrices, *Funct. Anal. Appl.* **24**(3), pp. 243–245.

Kosyak, A. V. (1992). Criteria for irreducibility and equivalence of regular Gaussian representations of group of finite upper triangular matrices of infinite order, *Sel. Math. Sov.* **11**, pp. 241–291.

Kosyak, A. V. (1994). Irreducible regular Gaussian representations of the group of the interval and the circle diffeomorphisms, *J. Funct. Anal.* **125**, pp. 493–547.

Kosyak, A. V. (2003). Irreducibility criterion for quasiregular representations of the group of finite uppertrianguler matrices, *Funct. Anal. Appl.* **37**(1), pp. 65–68.

Kosyak, A. V. (2004). Quasi-invariant measures and irreducible representations of the inductive limit of the special linear groups, *Funct. Anal. Appl.* **38**(1), pp. 67–68.

Kosyak, A. V. (2014). Induced representations of infinite-dimensional groups, I, *J. Funct. Anal.* **266**, pp. 3395–3434.

Kosyak, A. V. (2016). The Ismagilov conjecture over a finite field $\mathbb{F}_p$, arXiv:math. RT(GR)1612.01109.

Kosyak, A. V. (2018). Regular, quasi-regular and induced representations of infinite-dimensional groups, *EMS Tracts Math.* **29**, 587 p.

Kosyak, A. V. (2019). Criteria of irreducibility of the Koopman representations for the group $\mathrm{GL}_0(2\infty, \mathbb{R})$, *J. Funct. Anal.* **276**(1), pp. 78–126.

Kosyak, A. V. (2023). The generalized characteristic polynomial and corresponding resolvent with applications, arXiv:math.RT 2310.17351.

Kosyak, A. and Moree, P. (2025). Irreducible actions of the group $\mathrm{GL}(\infty)$ on L^2-spaces on 3 infinite rows, *J. Lie Theory* **35**(2), pp. 263–327.

Kosyak, A. V. (2025a). The height of an infinite parallelotope is infinite, *Linear Algebra Appl.* **709**, pp. 18–39, https://doi.org/10.1016/j.laa.2025.01.011.

Kuo, H. H. (1975). *Gaussian Measures in Banach Spaces*, Lecture Notes in Mathematics, Vol. 463, Springer, Berlin.

Lang, S. (1975). $SL_2(\mathbb{R})$, Addison-Wesley Publishing Co., Reading, Massachusetts.

Mackey, G. W. (1976). *The Theory of Unitary Group Representations*, The University of Chicago Press, Chicago, Ill.

Neeb, K. H. (2013). Unitary Representations of Unitary Groups, arXiv:math.RT. 1308.1500.

Neeb, K. H. (2017). Bounded and semi-bounded representations of infinite dimensional lie groups, in *Representation Theory—Current Trends and Perspectives*, P. Littelmann *et al.* (Eds.), EMS Series of Congress Reports, European Mathematical Society, Zürich, 2017, pp. 541–563, arXiv:math.RT:1510.08695.

Nessonov, N. I. (1986). Complete classification of representations of $GL(\infty)$ containing the identity representation of the unitary subgroup, *Mat. Sb. (N.S.)* **130**(2), pp. 242–262.

Ol'shanskiĭ, G. I. (1990). Unitary representations of (G, K)-pairs that are connected with the infinite symmetric group $S(\infty)$, *Algebra i Analiz.* **1**(4), pp. 178–209; *Leningrad Math. J.* **1**(4), pp. 983–1014.

Ol'shanskiĭ, G. I. (1990). Unitary representations of infinite-dimensional pairs (G, K) and the formalism of R. Howe, in: Representation of Lie Groups and Related Topics, A.M. Vershik and D.P. Zhelobenko (Eds.), *Advance Studies Contemporary Mathematics.* **7**, Gordon and Breach, New York, pp. 269–463.

Ol'shanskiĭ, G. I. (1991). Representations of infinite-dimensional classical groups, limits of enveloping algebras, and Yangians, in: Topics in Representation Theory, A.A. Kirillov (Ed.), *Advances in Soviet Mathematics,* Vol. 2, Amer. Math. Soc., Providence, RI, 1991, pp. 1–66.

Shilov, G. E. and Fan Dik Tun'. (1967). Integral, measure, and derivative on linear spaces (Russian), Nauka, Moscow.

Stratila, S and Voiculescu, D. (1975). *Representations of AF-algebras and of the group* $U(\infty)$, Lecture Notes in Mathematics, No. 486, Springer, New York.

Thoma, E. (1964). Die unzerlegbaren, positive-definiten Klassenfunktionen der abzählbar unendlichen, symmetrischen Gruppe, *Math. Z.* **85**, pp. 40–61.

Vershik, A. M. and Kerov, S. V. (1981). Asymptotic theory of characters of the symmetric group, *Funct. Anal. Appl.* **15**, pp. 246–255.

Vershik, A. M. and Kerov, S. V. (1981). Characters and factor representations of the infinite symmetric group, *Soviet Math. Dokl.* **23**, pp. 389–392.

Weil, A. (1953). *L'intégration dans les groupes topologique et ses application,* 2nd ed., Hermann, Paris.

Wolf, J. A. (2004). Principal Series Representations of Direct Limit Groups, arXiv:math/0402283 [math.RT].

Part 2
Representations of the Braid Groups

Chapter 11

Introduction

The *braid group* B_n was introduced by Artin (1926) to *study knots*:

$$B_n = \langle (\sigma_i)_{i=1}^{n-1} \mid \sigma_i \sigma_{i+1} \sigma_i = \sigma_{i+1} \sigma_i \sigma_{i+1}, \quad 1 \le i \le n-2,$$

$$\sigma_i \sigma_j = \sigma_i \sigma_j, \quad |\, i - j\,| \ge 2 \rangle. \tag{11.1}$$

Despite many profound results in the *classification of knots*, the general problem remains unsolved. We mention only a few polynomial invariants without any details: Alexander polynomials, Jones polynomials, Kauffman polynomials, Reshetikhin–Turaev quantum polynomials, Vassiliev invariants, etc. The same is true for the *classification of the representations* of B_n. See, e.g., Westbury (1995) for representations of B_3.

The classification of all irreducible representations of B_3 is equivalent to that of $\mathrm{PSL}(2, \mathbb{Z}) = \mathrm{SL}(2, \mathbb{Z})/\pm 1$. But $\mathrm{SL}(2, \mathbb{Z})$ contains the free group with two generators, F_2.

Our main aim is to connect two well-known representations of the braid groups: the Lawrence–Krammer (LK) and Burau representations of B_n.

In 1936, Burau introduced the family of representations of the braid group B_n as a *deformation* of the *standard representation* of the symmetric group S_n; see (12.1). The Burau representation for $n = 2, 3$ has been known to be *faithful* for some time, i.e., *the kernel of the corresponding representation is trivial.* Indeed, in 1969, Magnus and Peluso showed that the Burau and Gassner representations are faithful for $n = 2, 3$. See also Birman (1974, Theorem 3.15). Moody (1991), showed that the Burau representation is not faithful for $n \geq 9$; this result was improved to $n \geq 6$ by Long and Paton in 1992. The non-faithfulness for $n = 5$ was shown by Bigelow in 1999. The faithfulness of the Burau representation for $n = 4$ remains an *open problem.*

The Burau representation appears as a summand of the *Jones representation* (Jones, 1987), and for $n = 4$, the faithfulness of the Burau representation is equivalent to that of the Jones representation, which is furthermore related to the question of whether or not the Jones polynomial is an *unknot detector* (Bigelow, 2002b). The problem of *linearity*, i.e., the existence of the faithful representation, for the braid groups B_n for $n \geq 4$ remained open until 2000. In 1990, Lawrence constructed a two-parameter family of representations of the braid group B_n, using homological methods. In 2000, Krammer, using completely algebraic methods, constructed the same representations and showed that this representation is faithful for B_4. One year later, in 2001, Bigelow showed that the Lawrence–Krammer representation is faithful for all B_n. The following year, Krammer (2002) proved the same result using a different method. Finally, it was established that the braid groups B_n are linear for all n: in particular, for $n \leq 3$, one can use the Burau representation, and for $n \geq 4$, one can use the Lawrence–Krammer representation.

We show that the *Lawrence–Krammer representation is the quantisation of the symmetric square of the Burau representation.* A hint is the following: both representations of B_n have the same dimensions, $\frac{n(n-1)}{2}$, and their spectra coincide; see Remark 13.3. This connection allows us to construct new representations of the braid groups.

Three more hints are as follows: Lickorish (1989) observed that the Kauffman polynomial (related to the BMW algebras) at a particular parameter choice of $r = q^3$ is the same as the square of the Jones polynomial at some parameter choice. This paper investigates the relationships and interconnections between various polynomial invariants of classical links, particularly the Jones polynomial.

Zinno (2001) showed that the LK representation comes from the corresponding Birman–Murakami–Wenzl (BMW) algebra, i.e., for $r = q^3$ (his parametrisation is a little different). A connection is made between the LK representation and the BMW algebra. Inspired by a dimension argument, a basis is found for a certain irreducible representation of the algebra, and relations which generate the matrices are found.

Finally, Larsen and Rowell (2008) showed that the symmetric square of the Temperley–Lieb (TL) algebra was related to the same BMW algebra. More precisely, they established isomorphisms between certain specialisations of the BMW algebras and the symmetric squares of the TL algebras. Thus, the fact that the reduced Burau representation is a representation of the TL algebras and the LK representation is that of certain BMW algebras suggests that the main result of Kosyak (2025b) is a manifestation of this.

Formanek *et al.* (2003) also presented some classification results for all B_n. Namely, they classified all representations for B_n in dimension $\leq n$.

The *content* of the second part is as follows. In Chapter 12.1, we recall the definition of the Burau representation for B_n, and in Chapter 13, we define the Lawrence–Krammer representation $K_n^{(t,q)}$, with the corrections made by Bigelow. In Lemma 13.1, we establish the connection between $k_n^{(t,q)}$ and $k_{n+1}^{(t,q)}$; see (13.3). Some generalisations of the Burau representation are presented in Chapter 13.1. The main result, Theorem 13.1, is stated in Chapter 13.2. The *Pascal triangle* and its connection with representations of the group B_3 and with representations of a Lie algebra, $\mathfrak{sl}_2$, are described in Chapter 13.5. An important object, the *q-Pascal triangle*, and representations of B_3 (see Kosyak, 2008), connected to the Tuba–Wenzl results (Tuba, 2001; Tuba and Wenzl, 2001), are explained in Chapters 13.6–13.7. In Chapter 13.8, we define two symmetric bases of $V \otimes V$. The proof of Theorem 13.1 is given in Chapter 13.9. The essential ingredient is Lemma 13.1. Chapter 14.2 is devoted to the possible ways of constructing new representations of B_n, starting from representations of the Lie algebra $\mathfrak{sl}_{n-1}$.

Chapter 12

Connection Between the Lawrence–Krammer and the Burau Representations

12.1 The Burau Representation

The Burau representation $\rho : B_n \to \mathrm{GL}_n\big(\mathbb{Z}[t, t^{-1}]\big)$ is defined for a nonzero complex number t by

$$\sigma_i \mapsto I_{i-1} \oplus \begin{pmatrix} 1-t & t \\ 1 & 0 \end{pmatrix} \oplus I_{n-i-1}, \tag{12.1}$$

where $1-t$ is the (i, i)th entry. For $t = 1$, this is the *standard representation* of the symmetric group S_n, interchanging the two neighboring basis elements e_i and e_{i+1} in the space $\mathbb{C}^n$ via the matrix $\begin{pmatrix} 0 & 1 \\ 1 & 0 \end{pmatrix}$. The vector $e = e_1 + \cdots + e_n$ is invariant; therefore, the representation ρ splits into one-dimensional and $n - 1$−dimensional irreducible representations, known as the *reduced Burau representation*, $\rho_n^{(t)} : B_n \to \mathrm{GL}_{n-1}\big(\mathbb{Z}[t, t^{-1}]\big)$ and is defined by

$$\sigma_1 \mapsto \begin{pmatrix} -t & 0 \\ -1 & 1 \end{pmatrix} \oplus I_{n-3}, \quad \sigma_{n-1} \mapsto I_{n-3} \oplus \begin{pmatrix} 1 & -t \\ 0 & -t \end{pmatrix}, \tag{12.2}$$

$$\sigma_i \mapsto I_{i-2} \oplus \begin{pmatrix} 1 & -t & 0 \\ 0 & -t & 0 \\ 0 & -1 & 1 \end{pmatrix} \oplus I_{n-i-2}, \quad 2 \le i \le n - 2. \tag{12.3}$$

We use the following equivalent form of the reduced Burau representation: $\rho_n^{(t)} : B_n \mapsto \mathrm{GL}_{n-1}(\mathbb{Z}[t, t^{-1}])$, and we compare it with (12.2) and (12.3):

$$\sigma_1 \mapsto \begin{pmatrix} -t & t \\ 0 & 1 \end{pmatrix} \oplus I_{n-3}, \quad \sigma_{n-1} \mapsto I_{n-3} \oplus \begin{pmatrix} 1 & 0 \\ 1 & -t \end{pmatrix}, \tag{12.4}$$

$$\sigma_i \mapsto I_{i-2} \oplus \begin{pmatrix} 1 & 0 & 0 \\ 1 & -t & t \\ 0 & 0 & 1 \end{pmatrix} \oplus I_{n-i-2}, \quad 2 \leq i \leq n-2. \tag{12.5}$$

The equivalence is given by

$$J^{-1} \begin{pmatrix} 1 & -t & 0 \\ 0 & -t & 0 \\ 0 & -1 & 1 \end{pmatrix} J = \begin{pmatrix} 1 & 1 & 0 \\ 0 & -t & 0 \\ 0 & t & 1 \end{pmatrix} \overset{(\cdot)^T}{\mapsto} \begin{pmatrix} 1 & 0 & 0 \\ 1 & -t & t \\ 0 & 0 & 1 \end{pmatrix},$$

$$\text{where} \quad J = \begin{pmatrix} 0 & 0 & 1 \\ 0 & -1 & 0 \\ 1 & 0 & 0 \end{pmatrix},$$

and A^T is the matrix transposed to A.

12.2   The Reduced Burau Representation of B_∞

Define the group B_∞ as follows:

$$B_\infty = \langle (\sigma_i)_{i \in \mathbb{Z}} \mid \sigma_i \sigma_{i+1} \sigma_i = \sigma_{i+1} \sigma_i \sigma_{i+1}, \quad i \in \mathbb{Z},$$

$$\sigma_i \sigma_j = \sigma_i \sigma_j, \quad |i - j| \geq 2 \rangle. \tag{12.6}$$

Consider a complex Hilbert space:

$$H = l_2(\mathbb{Z}) = \left\{ x = (x_k)_{k \in \mathbb{Z}} \mid \|x\|^2 = \sum_{k \in \mathbb{Z}} |x_k|^2 < \infty \right\}, \tag{12.7}$$

with its standard orthonormal basis $e = (e_k)_{k \in \mathbb{Z}}$ defined by $e_k = (\delta_{kn})_{n \in \mathbb{Z}}$. Define the reduced Burau representation of the group B_∞ in the space H as follows:

$$\sigma_i \mapsto b_{i,\infty} = I_\infty \oplus \begin{pmatrix} 1 & 0 & 0 \\ 1 & -t & t \\ 0 & 0 & 1 \end{pmatrix} \oplus I_\infty, \quad i \in \mathbb{Z}, \qquad (12.8)$$

where $-t$ is the (i,i)th entry. It is clear that $b_{i+1,\infty}$ is obtained from $b_{i,\infty}$ by shifting the basis $e : e_n \to e_{n+1}$. More precisely, if we denote by U the unitary operator on $l_2(\mathbb{Z})$, defined as $Ue_n = e_{n+1}$, $n \in \mathbb{Z}$, we get the representation

$$U b_{i,\infty} U^{-1} = b_{i+1,\infty}. \qquad (12.9)$$

Next, denote by P_n the orthogonal projector of H on the subspace H_n *generated by the first n vectors*, i.e., $H_n = \langle e_k \mid 1 \leq k \leq n \rangle$. Then, we get the reduced Burau representation of B_{n+1}:

$$P_n b_{k,\infty} P_n = \rho^{(t)}_{n+1}(\sigma_k), \quad 1 \leq k \leq n. \qquad (12.10)$$

Denote by $J(n)$ the unitary operator on H_n, defined as follows

$$J(n)e_k = e_{k+1}, \quad 1 \leq k \leq n-1, \quad J(n)e_n = e_1, \qquad (12.11)$$

and by i_n, the natural embedding of $\mathrm{GL}(n-1,\mathbb{C})$ into $\mathrm{GL}(n,\mathbb{C})$ defined as follows:

$$\mathrm{GL}(n-1,\mathbb{C}) \ni x \to i_n(x) = x + E_{nn} \in \mathrm{GL}(n,\mathbb{C}). \qquad (12.12)$$

Then, we get for $n \geq 3$, $1 \leq k \leq n-2$ and $n-1 \leq r \leq n$,

$$\rho^{(t)}_{n+1}(\sigma_k) = i_n\big(\rho^{(t)}_n(\sigma_k)\big) \quad \text{and} \quad J(n)i_n\big(\rho^{(t)}_n(\sigma_k)\big)J(n)^{-1} = \rho^{(t)}_{n+1}(\sigma_{k+1}).$$

$$(12.13)$$

Remark 12.1. Formula (12.13) allows us to construct the reduced Burau representation for B_{n+1} from the reduced Burau representation for B_n. For $n = 4$, we have

$$
\rho_4^{(t)}(\sigma_1) = \begin{pmatrix} -t & t & 0 \\ 0 & 1 & 0 \\ 0 & 0 & 1 \end{pmatrix}, \quad
\rho_4^{(t)}(\sigma_2) = \begin{pmatrix} 1 & 0 & 0 \\ 1 & -t & t \\ 0 & 0 & 1 \end{pmatrix},
$$

$$
\rho_4^{(t)}(\sigma_3) = \begin{pmatrix} 1 & 0 & 0 \\ 0 & 1 & 0 \\ 0 & 1 & -t \end{pmatrix},
$$

For $n = 5$, we have

$$
\rho_5^{(t)}(\sigma_1) = \begin{pmatrix} -t & t & 0 & 0 \\ 0 & 1 & 0 & 0 \\ 0 & 0 & 1 & 0 \\ 0 & 0 & 0 & 1 \end{pmatrix}, \quad
\rho_5^{(t)}(\sigma_2) = \begin{pmatrix} 1 & 0 & 0 & 0 \\ 1 & -t & t & 0 \\ 0 & 0 & 1 & 0 \\ 0 & 0 & 0 & 1 \end{pmatrix},
$$

$$
\rho_5^{(t)}(\sigma_3) = \begin{pmatrix} 1 & 0 & 0 & 0 \\ 0 & 1 & 0 & 0 \\ 0 & 1 & -t & t \\ 0 & 0 & 0 & 1 \end{pmatrix}, \quad
\rho_5^{(t)}(\sigma_4) = \begin{pmatrix} 1 & 0 & 0 & 0 \\ 0 & 1 & 0 & 0 \\ 0 & 0 & 1 & 0 \\ 0 & 0 & 1 & -t \end{pmatrix}, \quad (12.14)
$$

so we get for $1 \le k \le 2$, and for $3 \le r \le 4$:

$$
\rho_5^{(t)}(\sigma_k) = i_n\big(\rho_4^{(t)}(\sigma_k)\big) \quad \text{and} \quad \rho_5^{(t)}(\sigma_r) = J(4)i_4\big(\rho_4^{(t)}(\sigma_{r-1})\big) J(4)^{-1}.
$$

$$
(12.15)
$$

$$\text{Chapter 13}$$

The Lawrence–Krammer Representation of the Braid Group B_n

The *Lawrence–Krammer representation* (LK) $K_n^{(t,q)}$ of the Braid group B_n was introduced by Lawrence (1990) and further studied by Krammer (2000, 2002) and Bigelow (2001, 2002a). To obtain formulas, we follow Bigelow (2001) and the corrected version in Bigelow (2002a). $K_n^{(t,q)}$ is the following action of B_n on a free module V of rank $\binom{n}{2} = \frac{n!}{2!(n-2)!}$ with basis $\{F_{i,j} : 1 \leq i < j \leq n\}$:

$$\sigma_i(F_{j,k}) = \begin{cases} F_{j,k}, & i \notin \{j-1, j, k-1, k\}, \\ qF_{i,k} + q(q-1)F_{i,j} + (1-q)F_{j,k}, & i = j-1, \\ F_{j+1,k}, & i = j \neq k-1, \\ qF_{j,i} + (1-q)F_{j,k} + q(1-q)tF_{i,k}, & i = k-1 \neq j, \\ F_{j,k+1}, & i = k, \\ -tq^2 F_{j,k}, & i = j = k-1. \end{cases}$$

$$(13.1)$$

From Bigelow (2002a), "This action is given in Krammer (2000), except that Krammer's $-t$ is my t. It is also in Bigelow (2001), but with a sign error. The name 'Krammer representation' was chosen because Krammer seems to have initially found this independently of Lawrence and without any use of homology".

We would like to modify the notation and define the *Lawrence–Krammer representation* $k_n^{(t,q)}$ as follows:

$$\sigma_i(F_{j,k}) = \begin{cases} F_{j,k}, & i \notin \{j-1, j, k-1, k\}, \\ tF_{i,k} + t(t-1)F_{i,j} + (1-t)F_{j,k}, & i = j-1, \\ F_{j+1,k}, & i = j \neq k-1, \\ tF_{j,i} + (1-t)F_{j,k} + t(t-1)qF_{i,k}, & i = k-1 \neq j, \\ F_{j,k+1}, & i = k, \\ qt^2 F_{j,k}, & i = j = k-1. \end{cases}$$

$$(13.2)$$

The connection between the two representations is as follows:

$$K_n^{(q,-t)}(\sigma_r) = k_n^{(t,q)}(\sigma_r), \quad 1 \le r \le n-1. \tag{13.3}$$

We believe that this choice of the parameters reflects better the situation since, historically, Burau used the parameter t, and q in our notation is in fact *a parameter of quantisation*; see (13.44). In particular, the LK representations $k_n^{(t,q)}$ with notation (13.2) for B_n, where $2 \le n \le 6$, are as follows[1]:

$$k_2^{(t,q)}(\sigma_1) = qt^2, \tag{13.4}$$

$$k_3^{(t,q)}(\sigma_1) = \left(\begin{array}{c|cc} qt^2 & 0 & t(t-1) \\ \hline 0 & 0 & t \\ 0 & 1 & 1-t \end{array}\right),$$

$$k_3^{(t,q)}(\sigma_2) = \left(\begin{array}{cc|c} 0 & t & 0 \\ 1 & 1-t & 0 \\ \hline 0 & qt(t-1) & qt^2 \end{array}\right), \tag{13.5}$$

---

[1]Together with Prof. Yang Hui-He and Dr. Mikhail Burtsev, both from LIMS, we found the formulas for B_5.

$$k_4^{(t,q)}(\sigma_1) = \begin{pmatrix} qt^2 & 0 & 0 & t(t-1) & t(t-1) & 0 \\ 0 & 0 & 0 & t & 0 & 0 \\ 0 & 0 & 0 & 0 & t & 0 \\ 0 & 1 & 0 & 1-t & 0 & 0 \\ 0 & 0 & 1 & 0 & 1-t & 0 \\ 0 & 0 & 0 & 0 & 0 & 1 \end{pmatrix},$$

$$k_4^{(t,q)}(\sigma_2) = \left(\begin{array}{ccc|ccc} 0 & t & 0 & 0 & 0 & 0 \\ 1 & 1-t & 0 & 0 & 0 & 0 \\ 0 & 0 & 1 & 0 & 0 & 0 \\ \hline 0 & qt(t-1) & 0 & qt^2 & 0 & t(t-1) \\ 0 & 0 & 0 & 0 & 0 & t \\ 0 & 0 & 0 & 0 & 1 & 1-t \end{array} \right),$$

$$k_4^{(t,q)}(\sigma_3) = \left(\begin{array}{ccc|ccc} 1 & 0 & 0 & 0 & 0 & 0 \\ 0 & 0 & t & 0 & 0 & 0 \\ 0 & 1 & 1-t & 0 & 0 & 0 \\ \hline 0 & 0 & 0 & 0 & t & 0 \\ 0 & 0 & 0 & 1 & 1-t & 0 \\ 0 & 0 & qt(t-1) & 0 & qt(t-1) & qt^2 \end{array} \right). \qquad (13.6)$$

$$k_5^{(t,q)}(\sigma_1) = \begin{pmatrix} qt^2 & 0 & 0 & 0 & t(t-1) & t(t-1) & t(t-1) & 0 & 0 & 0 \\ 0 & 0 & 0 & 0 & t & 0 & 0 & 0 & 0 & 0 \\ 0 & 0 & 0 & 0 & 0 & t & 0 & 0 & 0 & 0 \\ 0 & 0 & 0 & 0 & 0 & 0 & t & 0 & 0 & 0 \\ 0 & 1 & 0 & 0 & 1-t & 0 & 0 & t & 0 & 0 \\ 0 & 0 & 1 & 0 & 0 & 1-t & 0 & 0 & 0 & 0 \\ 0 & 0 & 0 & 1 & 0 & 0 & 1-t & 0 & 0 & 0 \\ 0 & 0 & 0 & 0 & 0 & 0 & 0 & 1 & 0 & 0 \\ 0 & 0 & 0 & 0 & 0 & 0 & 0 & 0 & 1 & 0 \\ 0 & 0 & 0 & 0 & 0 & 0 & 0 & 0 & 0 & 1 \end{pmatrix},$$

170   *Representations of Infinite-Dimensional and Braid Groups*

$$k_5^{(t,q)}(\sigma_2) = \left(\begin{array}{cccc|cccccc}
0 & t & 0 & 0 & 0 & 0 & 0 & 0 & 0 & 0 \\
1 & 1-t & 0 & 0 & 0 & 0 & 0 & 0 & 0 & 0 \\
0 & 0 & 1 & 0 & 0 & 0 & 0 & 0 & 0 & 0 \\
0 & 0 & 0 & 1 & 0 & 0 & 0 & 0 & 0 & 0 \\
\hline
0 & qt(t-1) & 0 & 0 & qt^2 & 0 & 0 & t(t-1) & t(t-1) & 0 \\
0 & 0 & 0 & 0 & 0 & 0 & 0 & t & 0 & 0 \\
0 & 0 & 0 & 0 & 0 & 0 & 0 & 0 & t & 0 \\
0 & 0 & 0 & 0 & 0 & 1 & 0 & 1-t & 0 & 0 \\
0 & 0 & 0 & 0 & 0 & 0 & 1 & 0 & 1-t & 0 \\
0 & 0 & 0 & 0 & 0 & 0 & 0 & 0 & 0 & 1
\end{array}\right),$$

$$k_5^{(t,q)}(\sigma_3) = \left(\begin{array}{cccc|cccccc}
1 & 0 & 0 & 0 & 0 & 0 & 0 & 0 & 0 & 0 \\
0 & 0 & t & 0 & 0 & 0 & 0 & 0 & 0 & 0 \\
0 & 1 & 1-t & 0 & 0 & 0 & 0 & 0 & 0 & 0 \\
0 & 0 & 0 & 1 & 0 & 0 & 0 & 0 & 0 & 0 \\
\hline
0 & 0 & 0 & 0 & 0 & t & 0 & 0 & 0 & 0 \\
0 & 0 & 0 & 0 & 1 & 1-t & 0 & 0 & 0 & 0 \\
0 & 0 & 0 & 0 & 0 & 0 & 1 & 0 & 0 & 0 \\
0 & 0 & qt(t-1) & 0 & 0 & qt(t-1) & 0 & qt^2 & 0 & t(t-1) \\
0 & 0 & 0 & 0 & 0 & 0 & 0 & 0 & 0 & t \\
0 & 0 & 0 & 0 & 0 & 0 & 0 & 0 & 1 & 1-t
\end{array}\right),$$

$$k_5^{(t,q)}(\sigma_4) = \left(\begin{array}{cccc|cccccc}
1 & 0 & 0 & 0 & 0 & 0 & 0 & 0 & 0 & 0 \\
0 & 1 & 0 & 0 & 0 & 0 & 0 & 0 & 0 & 0 \\
0 & 0 & 0 & t & 0 & 0 & 0 & 0 & 0 & 0 \\
0 & 0 & 1 & 1-t & 0 & 0 & 0 & 0 & 0 & 0 \\
\hline
0 & 0 & 0 & 0 & 1 & 0 & 0 & 0 & 0 & 0 \\
0 & 0 & 0 & 0 & 0 & 0 & t & 0 & 0 & 0 \\
0 & 0 & 0 & 0 & 0 & 1 & 1-t & 0 & 0 & 0 \\
0 & 0 & 0 & 0 & 0 & 0 & 0 & 0 & t & 0 \\
0 & 0 & 0 & 0 & 0 & 0 & 0 & 1 & 1-t & 0 \\
0 & 0 & 0 & qt(t-1) & 0 & 0 & qt(t-1) & 0 & qt(t-1) & qt^2
\end{array}\right).$$

$$(13.7)$$

The formulas of the LK representation $k_n^{(t,q)}(\sigma_r)$ for B_6 are as follows:

$$k_6^{(t,q)}(\sigma_1) = \begin{pmatrix}
qt^2 & 0 & 0 & 0 & 0 & t(t-1) & t(t-1) & t(t-1) & t(t-1) & 0 & 0 & 0 & 0 & 0 & 0 \\
0 & 0 & 0 & 0 & 0 & t & 0 & 0 & 0 & 0 & 0 & 0 & 0 & 0 & 0 \\
0 & 0 & 0 & 0 & 0 & 0 & t & 0 & 0 & 0 & 0 & 0 & 0 & 0 & 0 \\
0 & 0 & 0 & 0 & 0 & 0 & 0 & t & 0 & 0 & 0 & 0 & 0 & 0 & 0 \\
0 & 0 & 0 & 0 & 0 & 0 & 0 & 0 & t & 0 & 0 & 0 & 0 & 0 & 0 \\
0 & 1 & 0 & 0 & 0 & 1-t & 0 & 0 & 0 & 0 & 0 & 0 & 0 & 0 & 0 \\
0 & 0 & 1 & 0 & 0 & 0 & 1-t & 0 & 0 & 0 & 0 & 0 & 0 & 0 & 0 \\
0 & 0 & 0 & 1 & 0 & 0 & 0 & 1-t & 0 & 0 & 0 & 0 & 0 & 0 & 0 \\
0 & 0 & 0 & 0 & 1 & 0 & 0 & 0 & 1-t & 0 & 0 & 0 & 0 & 0 & 0 \\
0 & 0 & 0 & 0 & 0 & 0 & 1 & 0 & 0 & 1 & 0 & 0 & 0 & 0 & 0 \\
0 & 0 & 0 & 0 & 0 & 0 & 0 & 1 & 0 & 0 & 1 & 0 & 0 & 0 & 0 \\
0 & 0 & 0 & 0 & 0 & 0 & 0 & 0 & 0 & 0 & 0 & 1 & 0 & 0 & 0 \\
0 & 0 & 0 & 0 & 0 & 0 & 0 & 0 & 0 & 0 & 0 & 0 & 1 & 0 & 0 \\
0 & 0 & 0 & 0 & 0 & 0 & 0 & 0 & 0 & 0 & 0 & 0 & 0 & 1 & 0 \\
0 & 0 & 0 & 0 & 0 & 0 & 0 & 0 & 0 & 0 & 0 & 0 & 0 & 0 & 1
\end{pmatrix},$$

$$k_6^{(t,q)}(\sigma_2) = \left(\begin{array}{ccccc|cccccccccc}
0 & t & 0 & 0 & 0 & 0 & 0 & 0 & 0 & 0 & 0 & 0 & 0 & 0 & 0 \\
1 & 1-t & 0 & 0 & 0 & 0 & 0 & 0 & 0 & 0 & 0 & 0 & 0 & 0 & 0 \\
0 & 0 & 1 & 0 & 0 & 0 & 0 & 0 & 0 & 0 & 0 & 0 & 0 & 0 & 0 \\
0 & 0 & 0 & 1 & 0 & 0 & 0 & 0 & 0 & 0 & 0 & 0 & 0 & 0 & 0 \\
0 & 0 & 0 & 0 & 1 & 0 & 0 & 0 & 0 & 0 & 0 & 0 & 0 & 0 & 0 \\
\hline
0 & qt(t-1) & 0 & 0 & 0 & qt^2 & 0 & 0 & 0 & t(t-1) & t(t-1) & t(t-1) & 0 & 0 & 0 \\
0 & 0 & 0 & 0 & 0 & 0 & 0 & 0 & 0 & t & 0 & 0 & 0 & 0 & 0 \\
0 & 0 & 0 & 0 & 0 & 0 & 0 & 0 & 0 & 0 & t & 0 & 0 & 0 & 0 \\
0 & 0 & 0 & 0 & 0 & 0 & 0 & 0 & 0 & 0 & 0 & t & 0 & 0 & 0 \\
0 & 0 & 0 & 0 & 0 & 0 & 1 & 0 & 0 & 1-t & 0 & 0 & t & 0 & 0 \\
0 & 0 & 0 & 0 & 0 & 0 & 0 & 1 & 0 & 0 & 1-t & 0 & 0 & 0 & 0 \\
0 & 0 & 0 & 0 & 0 & 0 & 0 & 0 & 1 & 0 & 0 & 1-t & 0 & 0 & 0 \\
0 & 0 & 0 & 0 & 0 & 0 & 0 & 0 & 0 & 0 & 0 & 0 & 1 & 0 & 0 \\
0 & 0 & 0 & 0 & 0 & 0 & 0 & 0 & 0 & 0 & 0 & 0 & 0 & 1 & 0 \\
0 & 0 & 0 & 0 & 0 & 0 & 0 & 0 & 0 & 0 & 0 & 0 & 0 & 0 & 1
\end{array}\right),$$

$$
k_6^{(t,q)}(\sigma_3) = \left(\begin{array}{ccccc|cccccccccc}
1 & 0 & 0 & 0 & 0 & 0 & 0 & 0 & 0 & 0 & 0 & 0 & 0 & 0 & 0 \\
0 & 0 & t & 0 & 0 & 0 & 0 & 0 & 0 & 0 & 0 & 0 & 0 & 0 & 0 \\
0 & 1 & 1-t & 0 & 0 & 0 & 0 & 0 & 0 & 0 & 0 & 0 & 0 & 0 & 0 \\
0 & 0 & 0 & 1 & 0 & 0 & 0 & 0 & 0 & 0 & 0 & 0 & 0 & 0 & 0 \\
0 & 0 & 0 & 0 & 1 & 0 & 0 & 0 & 0 & 0 & 0 & 0 & 0 & 0 & 0 \\
0 & 0 & 0 & 0 & 0 & 0 & t & 0 & 0 & 0 & 0 & 0 & 0 & 0 & 0 \\
0 & 0 & 0 & 0 & 0 & 1 & 1-t & 0 & 0 & 0 & 0 & 0 & 0 & 0 & 0 \\
0 & 0 & 0 & 0 & 0 & 0 & 0 & 1 & 0 & 0 & 0 & 0 & 0 & 0 & 0 \\
0 & 0 & 0 & 0 & 0 & 0 & 0 & 0 & 1 & 0 & 0 & 0 & 0 & 0 & 0 \\
0 & 0 & qt(t-1) & 0 & 0 & 0 & qt(t-1) & 0 & 0 & qt^2 & 0 & 0 & t(t-1) & t(t-1) & 0 \\
0 & 0 & 0 & 0 & 0 & 0 & 0 & 0 & 0 & 0 & 0 & 0 & t & 0 & 0 \\
0 & 0 & 0 & 0 & 0 & 0 & 0 & 0 & 0 & 0 & 0 & 0 & 0 & t & 0 \\
0 & 0 & 0 & 0 & 0 & 0 & 0 & 0 & 0 & 1 & 0 & 0 & 1-t & 0 & 0 \\
0 & 0 & 0 & 0 & 0 & 0 & 0 & 0 & 0 & 0 & 1 & 0 & 0 & 1-t & 0 \\
0 & 0 & 0 & 0 & 0 & 0 & 0 & 0 & 0 & 0 & 0 & 0 & 0 & 0 & 1
\end{array}\right),
$$

$$
k_6^{(t,q)}(\sigma_4) = \left(\begin{array}{ccccc|cccccccccc}
1 & 0 & 0 & 0 & 0 & 0 & 0 & 0 & 0 & 0 & 0 & 0 & 0 & 0 & 0 \\
0 & 1 & 0 & 0 & 0 & 0 & 0 & 0 & 0 & 0 & 0 & 0 & 0 & 0 & 0 \\
0 & 0 & 0 & t & 0 & 0 & 0 & 0 & 0 & 0 & 0 & 0 & 0 & 0 & 0 \\
0 & 0 & 1 & 1-t & 0 & 0 & 0 & 0 & 0 & 0 & 0 & 0 & 0 & 0 & 0 \\
0 & 0 & 0 & 0 & 1 & 0 & 0 & 0 & 0 & 0 & 0 & 0 & 0 & 0 & 0 \\
0 & 0 & 0 & 0 & 0 & 1 & 0 & 0 & 0 & 0 & 0 & 0 & 0 & 0 & 0 \\
0 & 0 & 0 & 0 & 0 & 0 & 0 & t & 0 & 0 & 0 & 0 & 0 & 0 & 0 \\
0 & 0 & 0 & 0 & 0 & 0 & 1 & 1-t & 0 & 0 & 0 & 0 & 0 & 0 & 0 \\
0 & 0 & 0 & 0 & 0 & 0 & 0 & 0 & 1 & 0 & 0 & 0 & 0 & 0 & 0 \\
0 & 0 & 0 & 0 & 0 & 0 & 0 & 0 & 0 & 0 & t & 0 & 0 & 0 & 0 \\
0 & 0 & 0 & 0 & 0 & 0 & 0 & 0 & 0 & 1 & 1-t & 0 & 0 & 0 & 0 \\
0 & 0 & 0 & 0 & 0 & 0 & 0 & 0 & 0 & 0 & 0 & 1 & 0 & 0 & 0 \\
0 & 0 & 0 & qt(t-1) & 0 & 0 & 0 & qt(t-1) & 0 & 0 & qt(t-1) & 0 & qt^2 & 0 & t(t-1) \\
0 & 0 & 0 & 0 & 0 & 0 & 0 & 0 & 0 & 0 & 0 & 0 & 0 & 0 & t \\
0 & 0 & 0 & 0 & 0 & 0 & 0 & 0 & 0 & 0 & 0 & 0 & 0 & 1 & 1-t
\end{array}\right),
$$

$$k_6^{(t,q)}(\sigma_5) = \left(\begin{array}{cccccccccccccccc}
1 & 0 & 0 & 0 & 0 & 0 & 0 & 0 & 0 & 0 & 0 & 0 & 0 & 0 & 0 \\
0 & 1 & 0 & 0 & 0 & 0 & 0 & 0 & 0 & 0 & 0 & 0 & 0 & 0 & 0 \\
0 & 0 & 1 & 0 & 0 & 0 & 0 & 0 & 0 & 0 & 0 & 0 & 0 & 0 & 0 \\
0 & 0 & 0 & 0 & t & 0 & 0 & 0 & 0 & 0 & 0 & 0 & 0 & 0 & 0 \\
0 & 0 & 0 & 1 & 1-t & 0 & 0 & 0 & 0 & 0 & 0 & 0 & 0 & 0 & 0 \\
0 & 0 & 0 & 0 & 0 & 1 & 0 & 0 & 0 & 0 & 0 & 0 & 0 & 0 & 0 \\
0 & 0 & 0 & 0 & 0 & 0 & 1 & 0 & 0 & 0 & 0 & 0 & 0 & 0 & 0 \\
0 & 0 & 0 & 0 & 0 & 0 & 0 & 0 & t & 0 & 0 & 0 & 0 & 0 & 0 \\
0 & 0 & 0 & 0 & 0 & 0 & 0 & 1 & 1-t & 0 & 0 & 0 & 0 & 0 & 0 \\
0 & 0 & 0 & 0 & 0 & 0 & 0 & 0 & 0 & 1 & 0 & 0 & 0 & 0 & 0 \\
0 & 0 & 0 & 0 & 0 & 0 & 0 & 0 & 0 & 0 & 0 & t & 0 & 0 & 0 \\
0 & 0 & 0 & 0 & 0 & 0 & 0 & 0 & 0 & 0 & 1 & 1-t & 0 & 0 & 0 \\
0 & 0 & 0 & 0 & 0 & 0 & 0 & 0 & 0 & 0 & 0 & 0 & 0 & t & 0 \\
0 & 0 & 0 & 0 & 0 & 0 & 0 & 0 & 0 & 0 & 0 & 0 & 1 & 1-t & 0 \\
0 & 0 & 0 & 0 & qt(t-1) & 0 & 0 & 0 & qt(t-1) & 0 & 0 & qt(t-1) & 0 & qt(t-1) & qt^2
\end{array}\right). \tag{13.8}$$

To connect the representations $k_{n+1}^{(t,q)}$ and $k_n^{(t,q)}$, we introduce some notations. Define, for $\sigma \in S_n$, where S_n is a permutation group, and

$$\sigma = \begin{pmatrix} 1 & 2 & \ldots & n-1 & n \\ \sigma(1) & \sigma(2) & \ldots & \sigma(n-1) & \sigma(n) \end{pmatrix},$$

the *basis permutation operator* $s_\sigma \in \mathrm{GL}(n, \mathbb{C})$, given by

$$s_\sigma e_k = e_{\sigma(k)}, \quad 1 \leq k \leq n, \tag{13.9}$$

where $(e_k)_{k=1}^n$ is a standard basis in $\mathbb{C}^n$.

Lemma 13.1.[2] *We have the following relations between $k_n^{(t,q)}$ and $k_{n+1}^{(t,q)}$:*

$$k_{n+1}^{(t,q)}(\sigma_r) - qt(t-1)E_{(1,r+1),(r,r+1)} = D_{n+1,r} \oplus k_n^{(t,q)}(\sigma_{r-1}), \ \leq r \leq n,$$

$$s_{\sigma_{n+1,1}}\big(k_{n+1}^{(t,q)}(\sigma_1) - t(t-1)E_{(2,n+1),(1,2)}\big)s_{\sigma_{n+1,1}}^{-1} = k_n^{(t,q)}(\sigma_1) \oplus D_{n+1,1},$$

$$(13.10)$$

where $D_{n+1,r}$ and $s_{\sigma_{n+1,1}}$ are defined as follows.

Proof. If we compare (13.5) and (13.4), we get

$$k_3^{(t,q)}(\sigma_2) - qt(t-1)E_{13,23} = D_{32} \oplus k_2^{(t,q)}(\sigma_1), \qquad (13.11)$$

$$k_3^{(t,q)}(\sigma_1) - t(t-1)E_{23,12} = k_2^{(t,q)}(\sigma_1) \oplus D_{31}, \qquad (13.12)$$

$$\text{where } D_{32} = D_{31} = \begin{pmatrix} 0 & t \\ 1 & 1-t \end{pmatrix}, \ \sigma_{31} = \begin{pmatrix} 1 & 2 & 3 \\ 1 & 2 & 3 \end{pmatrix} = e \in S_3.$$

$$(13.13)$$

The expression for σ_{31} follows from the arrangement of the lexicographic order

$$\begin{pmatrix} 12 & 13 \\ & 23 \end{pmatrix}, \ \begin{pmatrix} 1 & 2 \\ & 3 \end{pmatrix} \rightarrow \begin{pmatrix} 1 & 2 \\ & 3 \end{pmatrix}. \qquad (13.14)$$

If we compare (13.5) and (13.6), we get

$$k_4^{(t,q)}(\sigma_3) - qt(t-1)E_{14,34} = D_{43} \oplus k_3^{(t,q)}(\sigma_2), \qquad (13.15)$$

$$k_4^{(t,q)}(\sigma_2) - qt(t-1)E_{13,23} = D_{42} \oplus k_3^{(t,q)}(\sigma_1), \qquad (13.16)$$

$$s_{\sigma_{41}}\big(k_4^{(t,q)}(\sigma_1) - t(t-1)E_{24,12}\big)s_{\sigma_{41}}^{-1} = k_3^{(t,q)}(\sigma_1) \oplus D_{41}, \quad (13.17)$$

$$\text{where } D_{43} = \begin{pmatrix} 1 & 0 & 0 \\ 0 & 0 & t \\ 0 & 1 & 1-t \end{pmatrix}, \ D_{42} = D_{41} = \begin{pmatrix} 0 & t & 0 \\ 1 & 1-t & 0 \\ 0 & 0 & 1 \end{pmatrix},$$

$$(13.18)$$

---

[2] After proving this lemma, the author asked Dr. Mikhail Burtsev from LIMS to find a connection between $k_n^{(t,q)}$ and $k_{n+1}^{(t,q)}$ with GPT o3. The result was not the same but rather similar.

$$\text{and } \sigma_{41} = \begin{pmatrix} 1 & 2| & 3 & 4 & |5 & 6 \\ 1 & 2| & 4 & 3 & |5 & 6 \end{pmatrix}. \tag{13.19}$$

The expression for σ_{41} follows from the arrangement of the lexicographic order

$$\begin{pmatrix} 12 & 13 & 14 \\ & 23 & 24 \\ & & 34 \end{pmatrix}, \quad \begin{pmatrix} 1 & 2 & 3 \\ & 4 & 5 \\ & & 6 \end{pmatrix} \rightarrow \begin{pmatrix} 1 & 2 & 4 \\ & 3 & 5 \\ & & 6 \end{pmatrix}. \tag{13.20}$$

If we compare (13.6) and (13.7), we can observe

$$k_5^{(t,q)}(\sigma_4) - qt(t-1)E_{15,45} = D_{54} \oplus k_4^{(t,q)}(\sigma_3), \tag{13.21}$$

$$k_5^{(t,q)}(\sigma_3) - qt(t-1)E_{14,34} = D_{53} \oplus k_3^{(t,q)}(\sigma_2), \tag{13.22}$$

$$k_5^{(t,q)}(\sigma_2) - qt(t-1)E_{13,23} = D_{52} \oplus k_3^{(t,q)}(\sigma_1), \tag{13.23}$$

$$s_{\sigma_{51}}\left(k_5^{(t,q)}(\sigma_1) - t(t-1)E_{25,12}\right)s_{\sigma_{51}}^{-1} = k_3^{(t,q)}(\sigma_1) \oplus D_{51}, \tag{13.24}$$

$$\text{where} \quad \sigma_{51} = \begin{pmatrix} 1 & 2 & 3| & 4 & 5 & 6 & 7 & 8 & |9 & 10 \\ 1 & 2 & 3| & 7 & 4 & 5 & 8 & 6 & |9 & 10 \end{pmatrix} \quad \text{and} \tag{13.25}$$

$$D_{54} = \left(\begin{array}{cc|cc} 1 & 0 & |0 & 0 \\ 0 & 1 & |0 & 0 \\ \hline 0 & 1 & |0 & t \\ 0 & 0 & |1 & 1-t \end{array}\right), \quad D_{53} = \left(\begin{array}{c|ccc} 1| & 0 & 0 & |0 \\ \hline 0| & 0 & t & |0 \\ 0| & 1 & 1-t & |0 \\ \hline 0| & 0 & 0 & |1 \end{array}\right),$$

$$D_{52} = D_{51} = \left(\begin{array}{cc|cc} 0 & t & |0 & 0 \\ 1 & 1-t & |0 & 0 \\ \hline 0 & 0 & |1 & 0 \\ 0 & 0 & |0 & 1 \end{array}\right). \tag{13.26}$$

The expression for σ_{51} follows from the arrangement of the lexicographic order

$$\begin{pmatrix} 12 & 13 & 14 & 15 \\ & 23 & 24 & 25 \\ & & 34 & 35 \\ & & & 45 \end{pmatrix}, \quad \begin{pmatrix} 1 & 2 & 3 & 4 \\ & 5 & 6 & 7 \\ & & 8 & 9 \\ & & & 10 \end{pmatrix} \rightarrow \begin{pmatrix} 1 & 2 & 3 & 7 \\ & 4 & 5 & 8 \\ & & 6 & 9 \\ & & & 10 \end{pmatrix}. \quad (13.27)$$

If we compare (13.7) and (13.8), we get the connection between $k_6^{(t,q)}$ and $k_5^{(t,q)}$:

$$k_6^{(t,q)}(\sigma_5) - qt(t-1)E_{16,56} = D_{65} \oplus k_5^{(t,q)}(\sigma_4), \qquad (13.28)$$

$$k_6^{(t,q)}(\sigma_4) - qt(t-1)E_{15,45} = D_{64} \oplus k_5^{(t,q)}(\sigma_3), \qquad (13.29)$$

$$k_6^{(t,q)}(\sigma_3) - qt(t-1)E_{14,34} = D_{63} \oplus k_5^{(t,q)}(\sigma_2), \qquad (13.30)$$

$$k_6^{(t,q)}(\sigma_2) - qt(t-1)E_{13,23} = D_{62} \oplus k_5^{(t,q)}(\sigma_1), \qquad (13.31)$$

$$s_{\sigma_{61}}\left(k_6^{(t,q)}(\sigma_1) - t(t-1)E_{26,12}\right)s_{\sigma_{61}}^{-1} = k_5^{(t,q)}(\sigma_1) \oplus D_{61}, \quad (13.32)$$

where
$$\sigma_{61} = \begin{pmatrix} 1\ 2\ 3\ 4| & 5 & 6\ 7\ 8 & 9 & 10\ 11 & 12\ 13 & |14\ 15 \\ 1\ 2\ 3\ 4| & 11 & 5\ 6\ 7 & 12 & 8 & 9 & 13\ 10 & |14\ 15 \end{pmatrix} \quad \text{and}$$

$$(13.33)$$

$$D_{65} = \left(\begin{array}{ccc|cc} 1 & 0 & 0 & 0 & 0 \\ 0 & 1 & 0 & 0 & 0 \\ 0 & 0 & 1 & 0 & 0 \\ \hline 0 & 0 & 1 & 0 & t \\ 0 & 0 & 0 & 1 & 1-t \end{array}\right), \quad D_{64} = \left(\begin{array}{cc|cc|c} 1 & 0 & 0 & 0 & 0 \\ 0 & 1 & 0 & 0 & 0 \\ \hline 0 & 0 & 0 & t & 0 \\ 0 & 0 & 1 & 1-t & 0 \\ \hline 0 & 0 & 0 & 0 & 1 \end{array}\right),$$

$$D_{63} = \left(\begin{array}{c|ccc|cc} 1 & 0 & 0 & 0 & 0 \\ \hline 0 & 0 & t & 0 & 0 \\ 0 & 1 & 1-t & 0 & 0 \\ \hline 0 & 0 & 0 & 1 & 0 \\ 0 & 0 & 0 & 0 & 1 \end{array}\right), \quad D_{62} = D_{61} = \left(\begin{array}{cc|ccc} 0 & t & 0 & 0 & 0 \\ 1 & 1-t & 0 & 0 & 0 \\ \hline 0 & 0 & 1 & 0 & 0 \\ 0 & 0 & 0 & 1 & 0 \\ 0 & 0 & 0 & 0 & 1 \end{array}\right).$$

$$(13.34)$$

The expression for σ_{61} follows from the arrangement of the lexico-graphic order

$$
\begin{pmatrix}
12 & 13 & 14 & 15 & 16 \\
 & 23 & 24 & 25 & 16 \\
 & & 34 & 35 & 36 \\
 & & & 45 & 46 \\
 & & & & 56
\end{pmatrix},
$$

$$
\begin{pmatrix}
1 & 2 & 3 & 4 & 5 \\
 & 6 & 7 & 8 & 9 \\
 & & 10 & 11 & 12 \\
 & & & 13 & 14 \\
 & & & & 15
\end{pmatrix}
\rightarrow
\begin{pmatrix}
1 & 2 & 3 & 4 & 11 \\
 & 5 & 6 & 7 & 12 \\
 & & 8 & 9 & 13 \\
 & & & 10 & 14 \\
 & & & & 15
\end{pmatrix}.
\tag{13.35}
$$

We can obtain the general relations (13.10) in a similar way. $\qquad\square$

Remark 13.1. We can observe that due to the description of the spectrum $k_n^{(t,q)}$, as given by (13.48), the diagonal part $\mathrm{diag}\big(k_n^{(t,q)}(\sigma_r)\big)$ of $k_n^{(t,q)}(\sigma_r)$ for all $1 \le r \le n$ are the same, up to some permutation:

$$
\mathrm{diag}\big(k_n^{t,q}(\sigma_r)\big) = \{qt^2, \underbrace{0,\ldots,0}_{n-2\ \text{times}}, \underbrace{1-t,\ldots,1-t}_{n-2\ \text{times}}, \underbrace{1,\ldots,1}_{(n-1)(n-2)/2-(n-2)\ \text{times}}\}.
\tag{13.36}
$$

13.1 Representations of B_{n+1} via Representations of the Lie Algebra $\mathfrak{gl}_n$

We can generalise the reduced Burau representation of B_n by simply noting (see Kosyak, 2008) that $\rho_n^{(t)}$ is a product of several exponentials of appropriate *Serre generators* of the *natural (or fundamental) representation* of the Lie algebra $\mathfrak{sl}_n$ (in fact, of $\mathfrak{gl}_n$); see (13.43). To be more precise, we show that the reduced Burau representation for B_3

$$\sigma_1 \mapsto \begin{pmatrix} -t & t \\ 0 & 1 \end{pmatrix}, \quad \sigma_2 \mapsto \begin{pmatrix} 1 & 0 \\ 1 & -t \end{pmatrix}, \tag{13.37}$$

can be rewritten as follows:

$$\sigma_1 \mapsto \begin{pmatrix} -t & t \\ 0 & 1 \end{pmatrix} = \begin{pmatrix} -t & 0 \\ 0 & 1 \end{pmatrix} \begin{pmatrix} 1 & -1 \\ 0 & 1 \end{pmatrix} = \exp(sE_{11}) \exp(-E_{12}),$$

$$\tag{13.38}$$

$$\sigma_2 \mapsto \begin{pmatrix} 1 & 0 \\ 1 & -t \end{pmatrix} = \begin{pmatrix} 1 & 0 \\ 1 & 1 \end{pmatrix} \begin{pmatrix} 1 & 0 \\ 0 & -t \end{pmatrix} = \exp(E_{21}) \exp(sE_{22}),$$

$$\tag{13.39}$$

where E_{kn} are *matrix unities* and $s = \ln(-t)$ for negative t.

For B_4, the reduced Burau representation is defined as follows:

$$\rho_4^{(t)} : B_4 \to \mathrm{GL}_3(\mathbb{Z}[t, t^{-1}]),$$

$$\sigma_1 \mapsto \begin{pmatrix} -t & t & 0 \\ 0 & 1 & 0 \\ 0 & 0 & 1 \end{pmatrix}, \ \sigma_2 \mapsto \begin{pmatrix} 1 & 0 & 0 \\ 1 & -t & t \\ 0 & 0 & 1 \end{pmatrix}, \ \sigma_3 \mapsto \begin{pmatrix} 1 & 0 & 0 \\ 0 & 1 & 0 \\ 0 & 1 & -t \end{pmatrix},$$

$$\tag{13.40}$$

$$\sigma_1 \mapsto \begin{pmatrix} -t & t & 0 \\ 0 & 1 & 0 \\ 0 & 0 & 1 \end{pmatrix} = \begin{pmatrix} -t & 0 & 0 \\ 0 & 1 & 0 \\ 0 & 0 & 1 \end{pmatrix} \begin{pmatrix} 1 & -1 & 0 \\ 0 & 1 & 0 \\ 0 & 0 & 1 \end{pmatrix}$$

$$= \exp(sE_{11}) \exp(-E_{12}),$$

$$\sigma_2 \mapsto \begin{pmatrix} 1 & 0 & 0 \\ 1 & -t & t \\ 0 & 0 & 1 \end{pmatrix} = \begin{pmatrix} 1 & 0 & 0 \\ 1 & 1 & 0 \\ 0 & 0 & 1 \end{pmatrix} \begin{pmatrix} 1 & 0 & 0 \\ 0 & -t & 0 \\ 0 & 0 & 1 \end{pmatrix} \begin{pmatrix} 1 & 0 & 0 \\ 0 & 1 & -1 \\ 0 & 0 & 1 \end{pmatrix}$$

$$= \exp(E_{21}) \exp(sE_{22}) \exp(-E_{23}),$$

$$\sigma_3 \mapsto \begin{pmatrix} 1 & 0 & 0 \\ 0 & 1 & 0 \\ 0 & 1 & -t \end{pmatrix} = \begin{pmatrix} 1 & 0 & 0 \\ 0 & 1 & 0 \\ 0 & 1 & 1 \end{pmatrix} \begin{pmatrix} 1 & 0 & 0 \\ 0 & 1 & 0 \\ 0 & 0 & -t \end{pmatrix}$$

$$= \exp(E_{32})\exp(sE_{33}),$$

where $s = \ln(-t)$ when $t < 0$. Finally, we get

$$\sigma_1 = \exp(sE_{11})\exp(-X_1), \quad \sigma_2 = \exp(Y_1)\exp(sE_{22})\exp(-X_2), \tag{13.41}$$

$$\sigma_3 = \exp(Y_2)\exp(sE_{33}). \tag{13.42}$$

Remark 13.2. In the general case for arbitrary B_n, the formulas are similar; see (13.43). We can use these formulas, i.e., the factorisation of σ_i, for the quantisation of the symmetric square of the reduced Burau representations, since it is easy to quantise the Pascal triangle; see Chapter 13.7, 13.9, especially formulas (13.92) and (13.94).

In what follows, we give the *Cartan matrix A*, corresponding to a Lie algebra $\mathfrak{sl}_n$, and "the rule to obtain" the right formulas, where $X_k = E_{kk+1}$, $Y_k = E_{k+1k}$ and $H_k = E_{kk} - E_{k+1,k+1}$, $1 \leq k \leq n-1$, *are the Serre generators of $\mathfrak{sl}_n$:*

$$A = \begin{pmatrix} 2 & -1 & 0 & \cdots & 0 & 0 & 0 \\ -1 & 2 & -1 & \cdots & 0 & 0 & 0 \\ 0 & -1 & 2 & \cdots & 0 & 0 & 0 \\ & & & \vdots & & & \\ 0 & 0 & 0 & \cdots & 2 & -1 & 0 \\ 0 & 0 & 0 & \cdots & -1 & 2 & -1 \\ 0 & 0 & 0 & \cdots & 0 & -1 & 2 \end{pmatrix},$$

$$
\begin{pmatrix}
E_{11} & -X_1 & & & & & \\
Y_1 & E_{22} & -X_2 & & & & \\
& Y_2 & E_{33} & -X_3 & & & \\
& & Y_3 & & & & \\
& & & & \ddots & & \\
& & & & & -X_{n-2} & \\
& & & & & E_{n-1,n-1} & -X_{n-1} \\
& & & & & Y_{n-1} & E_{nn}
\end{pmatrix}.
$$

We put the generators $X_k = E_{kk+1}$, $Y_k = E_{k+1k}$ in their appropriate places, corresponding to the indices of E_{kr}.

Lemma 13.2. *Let* $\pi : \mathfrak{gl}_n \to \mathrm{End}(V)$ *be an arbitrary representation of the Lie algebra* $\mathfrak{gl}_n$ *in the space* V. *Then, the formulas*

$$
\rho^{\pi}_{n+1}(\sigma_1) \mapsto \exp\big(s\pi(E_{11})\big)\exp\big(-\pi(X_1)\big),
$$

$$
\rho^{\pi}_{n+1}(\sigma_k) \mapsto \exp\big(\pi(Y_{k-1})\big)\exp\big(s\pi(E_{kk})\big)\exp(-\pi(X_k)),
$$

$$
\rho^{\pi}_{n+1}(\sigma_n) \mapsto \exp\big(\pi(Y_{n-1})\big)\exp\big(s\pi(E_{nn})\big) \tag{13.43}
$$

give us the representation $\rho^{\pi}_{n+1} : B_{n+1} \to \mathrm{GL}(V)$ *of the braid group* B_{n+1} *in the space* V. *With* π *being* the natural representation *of the Lie algebra* $\mathfrak{sl}_n$, *we get the reduced Burau representation* $\rho^{(t)}_{n+1}$.

Proof.   It is sufficient to use the fact that formulas (12.2) and (12.3) define the representation of the group B_n. $\square$

13.2   Connection Between the Lawrence–Krammer and Reduced Burau Representations

Theorem 13.1 (Kosyak, 2025b). *The LK representation* $k^{(t,q)}_n$ *defined by (13.2) is equivalent to the* quantisation *of the symmetric square of the reduced Burau representation, up to changing of the* *basis:*

$$k_n^{(t,q)} = s_{\sigma_{n1}}^{-1} \left[S^2(\rho_n^{(t)})_{e,v} \right]_q s_{\sigma_{n1}}; \qquad (13.44)$$

see Remark 13.11 *for an explicit explanation of the notation* $\left[S^2(\rho_n^{(t)})_{e,v} \right]_q$. *Here, the operator* s_σ *is defined by* (13.9)*, and* σ_{n1} *is defined in Lemma* 13.1*.*

Remark 13.3. In any representation of the group B_n, all generators σ_k, for $1 \leq k \leq n-1$, have the same spectrum since they are conjugated:

$$\sigma_k \sigma_{k+1} \sigma_k = \sigma_{k+1} \sigma_k \sigma_{k+1}. \qquad (13.45)$$

That is why we can speak about the *spectrum of the representation* of B_n.

The spectrum of the reduced Burau representation $\rho_n^{(t)}$ of B_n is as follows:

$$\operatorname{Sp} S^2(\rho_n^{(t)}(\sigma_k)) = \{t^2, \underbrace{-t, \ldots, -1,}_{n-2 \text{ times}} \underbrace{1, \ldots, 1}_{(n-1)(n-2)/2 \text{ times}} \}. \qquad (13.46)$$

The spectrum of its symmetric square is

$$\operatorname{Sp} S^2(\rho_n^{(t)}(\sigma_k)) = \{t^2, \underbrace{-t, \ldots, -t,}_{n-2 \text{ times}} \underbrace{1, \ldots, 1}_{(n-1)(n-2)/2 \text{ times}} \}. \qquad (13.47)$$

In Smeltzer (2003, Claim 5.8) (see also Stoimenov, 2008, Lemma 5), the following fact is proved.

Lemma 13.3. *The spectrum of the LK representation $LK := k_n^{(t,q)}$ of the group B_n, i.e., $\operatorname{Sp}\big(LK(\sigma_k)\big)$, is as follows:*

$$\operatorname{Sp}\big(LK(\sigma_k)\big) = \{qt^2, \underbrace{-t, \ldots, -t,}_{n-2 \text{ times}} \underbrace{1, \ldots, 1}_{(n-1)(n-2)/2 \text{ times}} \}. \qquad (13.48)$$

If we set $q = 1$ in (13.48), we get (13.47). Moreover, in Krammer (2000, Proposition 3.2), the following connection between the Krammer V module and the Burau W module was proved. Fix the representation $K_n^{(t,q)}$, defined by (13.1).

Proposition 1 (Krammer, 2000). *If $t = 1$ and 2 is invertible in R, then there is an isomorphism of B_n-modules $V \to S^2W$ (symmetric square of W) given by $v_{ij} \mapsto w_{ij}^2$ and, more generally, $v(T) \mapsto w(T)^2$.*

13.3   Summary

(1) We do not know how to quantise any representations of the groups B_n.

(2) But if the representation is an nth symmetric power of some representation B_n, then the Pascal triangle appears naturally; see Theorem 13.2, especially formula (13.57), we know how to quantise the Pascal triangle; see Chapter 13.6.

(3) The LK representation $k_n^{(t,q)}$, defined by (13.2) for $q = 1$, is the symmetric square of the Burau representation $S^2(\rho_n^{(t)})$ (see Proposition 1). That is why we can use the previous item.

(4) The fact that all irreducible representations of B_3 in $\dim V \leq 5$ obtained by Tuba and Wenzl [Tuba (2001); Tuba and Wenzl (2001)] can be given by formulas (13.59) (see Theorem 13.4), i.e., by the quantisation of the Pascal triangle (up to the diagonal matrix Λ_n), means that the q-Pascal triangle reflects the deep properties of the braid groups.

(5) Formulas (13.59) cover not only $\dim V \leq 5$, but in any dimension $n \in \mathbb{N}$, it also gives an $n(n-1)/2$-parameter family of representations of B_3.

13.4   Lie Algebra $\mathfrak{sl}_2$ and Its Finite-Dimensional Irreducible Representations

Recall that the Lie algebra $\mathfrak{sl}_2$ is the Lie algebra of 2×2 real matrices with its trace equal to zero. The standard basis X, Y, H in a Lie algebra $\mathfrak{sl}_2$ is as follows:

$$X = \begin{pmatrix} 0 & 1 \\ 0 & 0 \end{pmatrix}, \quad Y = \begin{pmatrix} 0 & 0 \\ 1 & 0 \end{pmatrix}, \quad H = \begin{pmatrix} 1 & 0 \\ 0 & -1 \end{pmatrix}. \tag{13.49}$$

These natural representations $\pi_2 : \mathfrak{sl}_2 \to \operatorname{End}(\mathbb{C}^2)$ are called *fundamental representations* of the Lie algebra $\mathfrak{sl}_2$. All finite-dimensional U-modules V that are the highest weight modules of highest weight λ are of the following form (see Kassel (1995, Theorem V.4.4.)):

$$\rho(n)(X) = S^n(\pi_2(X)) = \begin{pmatrix} 0 & n & 0 & \cdots & 0 \\ 0 & 0 & n-1 & \cdots & 0 \\ & & & \vdots & \\ 0 & 0 & 0 & \cdots & 1 \\ 0 & 0 & 0 & \cdots & 0 \end{pmatrix},$$

$$\rho(n)(Y) = S^n(\pi_2(Y)) = \begin{pmatrix} 0 & 0 & \cdots & 0 & 0 \\ 1 & 0 & \cdots & 0 & 0 \\ 0 & 2 & \cdots & 0 & 0 \\ & & \vdots & & \\ 0 & 0 & \cdots & n & 0 \end{pmatrix}, \tag{13.50}$$

$$\rho(n)(H) = S^n(\pi_2(H)) = \begin{pmatrix} n & 0 & \cdots & 0 & 0 \\ 0 & n-2 & \cdots & 0 & 0 \\ & & \vdots & & \\ & & \cdots & -n+2 & 0 \\ 0 & 0 & \cdots & 0 & -n \end{pmatrix}, \tag{13.51}$$

where $\lambda = \dim(V) - 1 \in \mathbb{N}$. In fact, all finite-dimensional irreducible representations of $\mathfrak{sl}_2$ are n-symmetric powers of the fundamental representation π_2. The main idea of the book by Kassel and Turaev (2008) is to give a comprehensive introduction to the theory of braid groups.

13.5   Pascal Triangle and the Representations of B_3 and Lie Algebra $\mathfrak{sl}_2$

The connection between the Pascal triangle and the representations of B_3 was first observed by Humphries (2000). Set in (13.59) $\Lambda_n = I$ and $q = 1$. Then,

$$\sigma_1 \mapsto \sigma_1(1, n), \quad \sigma_2 \mapsto \sigma_2(1, n), \quad \text{where} \quad \sigma_2(1, n) = \left(\sigma_1(1, n)^{-1}\right)^{\sharp} \tag{13.52}$$

is an irreducible representation of the group B_3 in $\mathbb{C}^{n+1}$. Here, the *Pascal triangle* denoted by $\sigma_1(1, n)$, for $n - 1 \in \mathbb{N}$, is defined as follows:

$$\sigma_1(1, 1) = \begin{pmatrix} 1 & 1 \\ 0 & 1 \end{pmatrix}, \quad \sigma_1(1, 2) = \begin{pmatrix} 1 & 2 & 1 \\ 0 & 1 & 1 \\ 0 & 0 & 1 \end{pmatrix},$$

$$\sigma_1(1, 3) = \begin{pmatrix} 1 & 3 & 3 & 1 \\ 0 & 1 & 2 & 1 \\ 0 & 0 & 1 & 1 \\ 0 & 0 & 0 & 1 \end{pmatrix}, \quad \text{and so forth,} \tag{13.53}$$

and *the central symmetry $A \mapsto A^{\sharp}$*, for the matrix $A \in \mathrm{Mat}(n + 1, \mathbb{C})$, $A = (a_{km})_{0 \leq k, m \leq n}$, is defined as follows:

$$A^{\sharp} = \left(a_{km}^{\sharp}\right)_{0 \leq k, m \leq n}, \quad a_{km}^{\sharp} = a_{n-k, n-m}. \tag{13.54}$$

Theorem 13.2. *Let π_2 be the fundamental representation of $\mathfrak{sl}_2$. Then,*

$$\sigma_1(1, n) = \exp\left(\rho(X)\right) = \exp\left(S^n(\pi_2(X))\right), \tag{13.55}$$

$$\sigma_2(1, n) = \exp\left(\rho(-Y)\right) = \exp\left(S^n(\pi_2(-Y))\right), \tag{13.56}$$

i.e., the Humphries representation of B_3 in dimension $n + 1$ is the nth symmetric power of the Burau representation $\rho_3^{(-1)}$.

As a result of (13.55), we obtain an elegant formula, $\begin{pmatrix} 1 & 3 & 3 & 1 \\ 0 & 1 & 2 & 1 \\ 0 & 0 & 1 & 1 \\ 0 & 0 & 0 & 1 \end{pmatrix} =$ $\exp\begin{pmatrix} 0 & 3 & 0 & 0 \\ 0 & 0 & 2 & 0 \\ 0 & 0 & 0 & 1 \\ 0 & 0 & 0 & 0 \end{pmatrix}$, and in the general case,

$$\sigma_1(1,n) = \begin{pmatrix} \binom{n}{n} & \binom{n}{n-1} & \binom{n}{n-2} & \cdots & \binom{n}{1} & \binom{n}{0} \\ 0 & \binom{n-1}{n-1} & \binom{n-1}{n-2} & \cdots & \binom{n-1}{1} & \binom{n-1}{0} \\ 0 & 0 & \binom{n-2}{n-2} & \cdots & \binom{n-2}{1} & \binom{n-2}{0} \\ & & & \vdots & & \\ 0 & 0 & 0 & \cdots & \binom{1}{1} & \binom{1}{0} \\ 0 & 0 & 0 & \cdots & 0 & \binom{0}{0} \end{pmatrix}$$

$$= \exp\left(\sum_{k=0}^{n-1} (n-k) E_{k,k+1} \right). \tag{13.57}$$

13.6 q-Pascal Triangle and Representations of B_3

We define the family of representations of the group B_3 using the q-*Pascal triangle* (Albeverio and Kosyak, 2007; Kosyak, 2008). For $q \in \mathbb{C}^{\times} := \mathbb{C} \setminus \{0\}$ and a matrix

$$\Lambda_n = \operatorname{diag}(\lambda_r)_{r=0}^n \quad \text{such that} \quad \lambda_r \lambda_{n-r} = const, \tag{13.58}$$

set

$$\sigma_1^{\Lambda_n}(q,n) := \sigma_1(q,n) D_n^{\sharp}(q)\Lambda_n, \quad \sigma_2^{\Lambda_n}(q,n) := \Lambda_n^{\sharp} D_n(q)\sigma_2(q,n), \tag{13.59}$$

$$D_n(q) = \operatorname{diag}(q_r)_{r=0}^n, \quad q_r = q^{\frac{(r-1)r}{2}}, \tag{13.60}$$

and $\sigma_1(q,n)$ is the q-*Pascal triangle* obtained from the Pascal triangle by replacing the natural numbers n with q-natural numbers $(n)_q$; see (13.65). More precisely, define $\sigma_1(q,n)$ as follows: $\sigma_1(q,n) = \big(\sigma_1(q,n)_{km}\big)_{0 \leq k,m \leq n}$, where

$$\sigma_1(q,n)_{km} = \binom{n-k}{n-m}_q, \quad 0 \leq k, m \leq n, \tag{13.61}$$

$$\sigma_2(q,n) := \big(\sigma_1^{-1}(q^{-1},n)\big)^{\sharp}, \tag{13.62}$$

$$\big(\sigma_1^{-1}(q^{-1},n)\big)^{\sharp}_{km} = \begin{cases} 0, & \text{if } 0 < k < m \le n, \\ (-1)^{k+m} q^{-1}_{k-m} C_k^m(q^{-1}), & \text{if } 0 \le m \le k \le n. \end{cases} \tag{13.63}$$

Here, the *q-binomial coefficients* or *Gaussian polynomials* are defined as follows:

$$C_n^k(q) := \binom{n}{k}_q := \frac{(n)!_q}{(k)!_q (n-k)!_q}, \quad C_n^k[q] := \begin{bmatrix} n \\ k \end{bmatrix}_q := \frac{[n]!_q}{[k]!_q [n-k]!_q}, \tag{13.64}$$

corresponding to the two forms of $q-natural\ numbers$, defined by

$$(n)_q := \frac{q^n - 1}{q - 1}, \quad [n]_q := \frac{q^n - q^{-n}}{q - q^{-1}}. \tag{13.65}$$

Remark 13.4. We have

$$D_n(q)\sigma_2(q,n) = j_n\big(\sigma_1(q,n)\big)^{\sharp} j_n^{-1} D_n(q), \quad \text{where} \quad j_n = \sum_{r=0}^{n}(-1)^r E_{r,n-r}. \tag{13.66}$$

Another way to obtain (13.59) is as follows. Define the *q-Pochhammer symbol*:

$$(a;q)_n = \prod_{k=0}^{n-1}(1-aq^k) = (1-a)(1-aq)(1-aq^2)\cdots(1-aq^{n-1}). \tag{13.67}$$

We have, referring to Andrews (1976),

$$(1+x)_q^k := (-x;q)_k = \sum_{r=0}^{k} q^{r(r-1)/2} C_k^r(q) x^r = \sum_{r=0}^{k} q^{r(r-1)/2} \left(\begin{smallmatrix} k \\ r \end{smallmatrix}\right)_q x^r. \tag{13.68}$$

Therefore, the inversion of the matrix $\sigma_1^{\Lambda_n}(q,n)$ for $\Lambda_n = I_{n+1}$ with respect to the auxiliary diagonal is the matrix of transformations of the monomials $x^0, x^1, x^2, \ldots, x^n$ under rule (13.68).

Theorem 13.3 (Albeverio and Kosyak, 2007; Kosyak, 2008).
The formulas $\sigma_1 \mapsto \sigma_1^{\Lambda_n}(q, n), \quad \sigma_2 \mapsto \sigma_2^{\Lambda_n}(q, n)$, referring to (13.59), define the family of representations of the group B_3 in dimension $n+1$.

Remark 13.5. Let $\pi : B_3 \to \mathrm{GL}(m, \mathbb{C})$ be a representation. Fix two matrices, $\Lambda_1, \Lambda_2 \in \mathrm{GL}(m, \mathbb{C})$, and define $\pi^{\Lambda}(\sigma_1) = \Lambda_1 \pi(\sigma_1), \quad \pi^{\Lambda}(\sigma_2) = \pi(\sigma_2)\Lambda_2$. When π^{Λ} is again a representation of B_3? Under the additional conditions $\Lambda_1 \Lambda_2 = cI$, $c \in \mathbb{C}$ and $\Lambda_1 = \mathrm{diag}(\lambda_k)_{k=0}^{n}$, the criterion for π^{Λ} to be a representation is the following:

$$\Lambda_1 \pi(\sigma_1 \sigma_2 \sigma_1) = \pi(\sigma_1 \sigma_2 \sigma_1)\Lambda_2. \tag{13.69}$$

This implies condition (13.58), i.e., $\lambda_r \lambda_{n-r} = c$.

Theorem 13.4 (Tuba, 2001; Kosyak, 2008). *All representations of the group B_3 in dimension ≤ 5 are given by the formulas (13.59).*

In dimension 2, we get for $\Lambda_1 = \mathrm{diag}(\lambda_0, \lambda_1)$,

$$\sigma_1^{\Lambda_1}(q, 1) = \begin{pmatrix} 1 & 1 \\ 0 & 1 \end{pmatrix}\begin{pmatrix} \lambda_0 & 0 \\ 0 & \lambda_1 \end{pmatrix}, \quad \sigma_2^{\Lambda_1}(q, 1) = \begin{pmatrix} \lambda_1 & 0 \\ 0 & \lambda_0 \end{pmatrix}\begin{pmatrix} 1 & 0 \\ -1 & 1 \end{pmatrix}. \tag{13.70}$$

In dimension 3, we get for $\Lambda_2 = \mathrm{diag}(\lambda_0, \lambda_1, \lambda_2)$, with $\lambda_0 \lambda_2 = \lambda_1^2$,

$$\sigma_1^{\Lambda_2}(q, 2) = \begin{pmatrix} 1 & (1+q) & 1 \\ 0 & 1 & 1 \\ 0 & 0 & 1 \end{pmatrix} D_2^{\sharp}(q)\Lambda_2,$$

$$\sigma_2^{\Lambda_2}(q, 2) = \Lambda_2^{\sharp} \begin{pmatrix} 1 & 0 & 0 \\ -1 & 1 & 0 \\ 1 & -(1+q) & 1 \end{pmatrix} D_2(q). \tag{13.71}$$

In dimension 4, we get for $\Lambda_3 = \mathrm{diag}(\lambda_0, \lambda_1, \lambda_2, \lambda_3)$, with $\lambda_0 \lambda_3 = \lambda_1 \lambda_2$, referring to Remark 13.5,

$$\sigma_1^{\Lambda_3}(q, 3) = \begin{pmatrix} 1 & (1+q+q^2) & (1+q+q^2) & 1 \\ 0 & 1 & (1+q) & 1 \\ 0 & 0 & 1 & 1 \\ 0 & 0 & 0 & 1 \end{pmatrix} D_3^{\sharp}(q)\Lambda_3, \tag{13.72}$$

$$\sigma_2^{\Lambda_3}(q,3) = \Lambda_3^{\sharp} \begin{pmatrix} 1 & 0 & 0 & 0 \\ -1 & 1 & 0 & 0 \\ 1 & -(1+q) & 1 & 0 \\ -1 & (1+q+q^2) & -(1+q+q^2) & 1 \end{pmatrix} D_3(q).$$

$$(13.73)$$

In dimension 5, we get for $\Lambda_4 = \mathrm{diag}(\lambda_r)_{r=0}^4$, with $\lambda_0\lambda_4 = \lambda_1\lambda_3 = \lambda_2^2$,

$\sigma_1^{\Lambda_4}(q,4)$

$$= \begin{pmatrix} 1 & (1+q)(1+q^2) & (1+q^2)(1+q+q^2) & (1+q)(1+q^2) & 1 \\ 0 & 1 & 1+q+q^2 & 1+q+q^2 & 1 \\ 0 & 0 & 1 & 1+q & 1 \\ 0 & 0 & 0 & 1 & 1 \\ 0 & 0 & 0 & 0 & 1 \end{pmatrix} D_4^{\sharp}(q)\Lambda_4,$$

$$(13.74)$$

$\sigma_2^{\Lambda_4}(q,4)$

$$= \Lambda_4^{\sharp} \begin{pmatrix} 1 & 0 & 0 & 0 & 0 \\ -1 & 1 & 0 & 0 & 0 \\ 1 & -(1+q) & 1 & 0 & 0 \\ -1 & (1+q+q^2) & -(1+q+q^2) & 1 & 0 \\ 1 & -(1+q)(1+q^2) & (1+q^2)(1+q+q^2) & -(1+q)(1+q^2) & 1 \end{pmatrix} D_4(q).$$

$$(13.75)$$

13.7   Quantisation of the Pascal Triangle

Remark 13.6. The following procedure of quantisation ("Quant") is based on the *quantisation of the Pascal triangle* explained in Albeverio and Kosyak (2007) and Kosyak (2008), given by formulas (10) and (11). To be more precise, we change the binomial coefficients $\binom{n}{k}$ of the Pascal triangle to the *q-binomial coefficients* $\binom{n}{k}_q = \frac{(n)!_q}{(k)!_q(n-k)!_q}$.

We introduce the representation $T_{3,n+1}^{\Lambda_n,q}$ of the braid group B_3, equivalent to the representation (13.59) in dimension $n+1$ as follows:

$$T_{3,n+1}^{\Lambda_n,q}(\sigma_1) = \left(\sigma_2^{\Lambda_n}(q,n)\right)^{\sharp}, \quad T_{3,n+1}^{\Lambda_n,q}(\sigma_2) = \left(\sigma_1^{\Lambda_n}(q,n)\right)^{\sharp}. \tag{13.76}$$

Lemma 13.4. *We have*

$$T_{3,n+1}^{\Lambda_n,q}(\sigma_1) = J_n \sigma_2^{\Lambda_n}(q,n) J_n^{-1}, \quad T_{3,n+1}^{\Lambda_n,q}(\sigma_2) = J_n \sigma_1^{\Lambda_n}(q,n) J_n^{-1}. \tag{13.77}$$

Proof. We use the fact that $A^{\sharp} = J_n A J_n^{-1}$, where $J_n = \sum_{r=0}^{n} E_{r,n-r}$. $\qquad\square$

Remark 13.7. Using Tuba and Wenzl (2001) (see also Kosyak, 2008), we conclude that all representations of B_3 in dimensions $2, 3, 4$ and 5 are equivalent to the following ones, where $\Lambda_1 = s\,\mathrm{diag}(-t,1)$ and $s, t \in \mathbb{C}^{\times}$:

$$T_{32}^{\Lambda_1,1,\pi_2}(\sigma_1) = s \begin{pmatrix} -t & t \\ 0 & 1 \end{pmatrix} = s \begin{pmatrix} -t & 0 \\ 0 & 1 \end{pmatrix} \begin{pmatrix} 1 & -1 \\ 0 & 1 \end{pmatrix},$$

$$T_{32}^{\Lambda_1,1,\pi_2}(\sigma_2) = s \begin{pmatrix} 1 & 0 \\ 1 & -t \end{pmatrix} = s \begin{pmatrix} 1 & 0 \\ 1 & 1 \end{pmatrix} \begin{pmatrix} 1 & 0 \\ 0 & -t \end{pmatrix}. \tag{13.78}$$

All representations of B_3 in dimension 3 are given by the following formulas, where $s\Lambda_2 = s\,\mathrm{diag}(t^2,-t,1) = s\,\mathrm{diag}(t^2,-t,1)$ and $s, t, q \in \mathbb{C}^{\times}$:

$$T_{33}^{\Lambda_2,q,\pi_2}(\sigma_1) = s \begin{pmatrix} t^2q & -t^2(1+q) & t^2 \\ 0 & -t & t \\ 0 & 0 & 1 \end{pmatrix}$$

$$= s\Lambda_2 \begin{pmatrix} 1 & -(1+q) & 1 \\ 0 & 1 & -1 \\ 0 & 0 & 1 \end{pmatrix} D_2^{\sharp}(q), \tag{13.79}$$

$$T_{33}^{\Lambda_2,q,\pi_2}(\sigma_2) = s \begin{pmatrix} 1 & 0 & 0 \\ 1 & -t & 0 \\ 1 & -t(1+q) & t^2 q \end{pmatrix} = s \begin{pmatrix} 1 & 0 & 0 \\ 1 & 1 & 0 \\ 1 & (1+q) & 1 \end{pmatrix} D_2(q)\Lambda_2^\sharp.$$

$$(13.80)$$

Remark 13.8. If we are sufficiently attentive, we can recognise here the LK representation of B_3 for $s = 1$; see formulas (13.95) and (13.102) later.

All representations of B_3 in dimension 4 are given by the following formulas, where $\Lambda_3 = s \operatorname{diag}\left(-t^3, ut^2, -u^{-1}t, t\right)$:

$$T_{34}^{\Lambda_3,q}(\sigma_1) = \Lambda_3 \begin{pmatrix} 1 & -(1+q+q^2) & (1+q+q^2) & -1 \\ 0 & 1 & -(1+q) & 1 \\ 0 & 0 & 1 & -1 \\ 0 & 0 & 0 & 1 \end{pmatrix} D_3^\sharp(q),$$

$$(13.81)$$

$$T_{34}^{\Lambda_3,q}(\sigma_2) = \begin{pmatrix} 1 & 0 & 0 & 0 \\ 1 & 1 & 0 & 0 \\ 1 & (1+q) & 1 & 0 \\ 1 & (1+q+q^2) & (1+q+q^2) & 1 \end{pmatrix} D_3(q)\Lambda_3^\sharp. \qquad (13.82)$$

In dimension 5, we get for $\Lambda_4 = \operatorname{diag}(\lambda_r)_{r=0}^4$, with $\lambda_0\lambda_4 = \lambda_1\lambda_3 = \lambda_2^2$,

$$T_{35}^{\Lambda_4,q}(\sigma_1)$$

$$= \Lambda_4 \begin{pmatrix} 1 & -(1+q)(1+q^2) & (1+q^2)(1+q+q^2) & -(1+q)(1+q^2) & -1 \\ 0 & 1 & -(1+q+q^2) & (1+q+q^2) & -1 \\ 0 & 0 & 1 & -(1+q) & 1 \\ 0 & 0 & 0 & 1 & -1 \\ 0 & 0 & 0 & 0 & 1 \end{pmatrix} D_4^\sharp(q),$$

$$(13.83)$$

$$T_{35}^{\Lambda_4,q}(\sigma_2)$$

$$= \begin{pmatrix} 1 & 0 & 0 & 0 & 0 \\ 1 & 1 & 0 & 0 & 0 \\ 1 & (1+q) & 1 & 0 & 0 \\ 1 & (1+q+q^2) & (1+q+q^2) & 1 & 0 \\ 1 & (1+q)(1+q^2) & (1+q^2)(1+q+q^2) & (1+q)(1+q^2) & 1 \end{pmatrix} D_4(q)\Lambda_4^\sharp.$$

$$(13.84)$$

Remark 13.9. The irreducibility criteria for the representation (13.59) of B_3 in dimensions $n \leq 5$ were obtained in Tuba (2001) and Tuba and Wenzl (2001).

Remark 13.10. For dimensions $n \geq 6$, formulas (13.59) provide us with a few families of representations of B_3 but not all; see Tuba and Wenzl (2001, Remark 2.11.3). For more representations of B_3, see, e.g., Westbury (1995). The question of irreducibility of representations (13.59) remains open for $n \geq 6$.

13.8 Symmetric Basis in $V \otimes V$

Let V be a finite-dimensional linear space with the basis $(e_k)_{k=1}^n$. The *tensor product* $V \otimes V$ is a linear space generated by the basis $\big(e_k \otimes e_r\big)_{k,r=1}^n$. The *symmetric square* $S^2(V)$ of the space V is a subspace of $V \otimes V$ generated by one of the equivalent *symmetric basis* $e = \big(e_{kn}^s\big)_{kn}$ or $v = \big(v_{kn}\big)_{kn}$, defined as follows:

$$e_{kk}^s = e_k \otimes e_k, \quad 1 \leq k \leq n, \quad e_{kr}^s = e_k \otimes e_r + e_r \otimes e_k, \quad 1 \leq k < r \leq n,$$

$$(13.85)$$

$$v_{kr} = (e_k + \cdots + e_{r-1}) \otimes (e_k + \cdots + e_{r-1}), \quad 1 \leq k < r \leq n.$$

$$(13.86)$$

We fix a *lexicographic order* on the set (k,r), with $1 \leq k \leq r \leq n$. Let an operator A acts on a space V. Denote by $A \otimes A$ its tensor product on a space $V \otimes V$ and by $S^2(A)$ its *symmetric square*.

13.9   Proof of Theorem 13.1

Remark 13.11. We show that $\left[S^2(\rho_n^{(t)})_{e,v}\right]_q$, the quantisation $[\ldots]_q$ of the symmetric square $S^2(\rho_n^{(t)})_e$ of the reduced Burau representations $\rho_n^{(t)}$ for the group B_n, expressed in the basis e defined by (13.85) and *calculated further in the basis* v defined by (13.86), coincide with the LK representation $k_n^{(t,q)}$ in the notation (13.2), i.e., we show that

$$s_{\sigma_{n1}}^{-1} C_n \left[S^2\left(\rho_n^{(t)}(\sigma_r)\right)_e\right]_q C_n^{-1} s_{\sigma_{n1}} = k_n^{(t,q)}(\sigma_r), \quad \text{for} \quad 1 \le r \le n-1,$$
$$(13.87)$$

where C_n is the *change-of-basis matrix* from the basis e, defined by (13.85), to the basis v, defined by (13.86), for the vector space $S^2(\mathbb{C}^{n-1})$. Here, the operator s_σ is defined by (13.9), and σ_{n1} is defined in Lemma 13.1.

The Burau representation $\rho_3^{(t)}$ for B_3 on the space $V = \mathbb{C}^2$ is as follows:

$$\sigma_1 \mapsto \begin{pmatrix} -t & t \\ 0 & 1 \end{pmatrix}, \quad \sigma_2 \mapsto \begin{pmatrix} 1 & 0 \\ 1 & -t \end{pmatrix}. \tag{13.88}$$

We fix the standard basis as $e = (e_k)_{k=1}^n$ in $\mathbb{C}^n$, namely set

$$e_k = (\underbrace{0, \ldots, 0, 1, 0, \ldots, 0}_{k \text{ place}}), \quad 1 \le k \le n. \tag{13.89}$$

In the symmetric square $S^2(\mathbb{C}^2)$, the bases e and v are the following:

$$e_{11}^s = e_1 \otimes e_1, \quad e_{12}^s = e_1 \otimes e_2 + e_2 \otimes e_1, \quad e_{22}^s = e_2 \otimes e_2, \tag{13.90}$$

$$v_{12} = e_1 \otimes e_1, \quad v_{13}^s = (e_1 + e_2) \otimes (e_1 + e_2), \quad v_{23} = e_2 \otimes e_2. \tag{13.91}$$

The *quantisation* of the symmetric square $\left[S^2(\sigma_k)_e\right]_q :=$ $\left[S^2(\rho_3^{(t)}(\sigma_k))_e\right]_q$ for $k = 1, 2$ of the Burau representation $\rho_3^{(t)}$ has the following form in the basis e:

$$\sigma_1 \mapsto \begin{pmatrix} -t & 0 \\ 0 & 1 \end{pmatrix}\begin{pmatrix} 1 & -1 \\ 0 & 1 \end{pmatrix} \overset{\text{Sym}^2}{\mapsto} \begin{pmatrix} t^2 & 0 & 0 \\ 0 & -t & 0 \\ 0 & 0 & 1 \end{pmatrix}\begin{pmatrix} 1 & -2 & 1 \\ 0 & 1 & -1 \\ 0 & 0 & 1 \end{pmatrix} \overset{\text{Quant}}{\mapsto}$$

$$\begin{pmatrix} t^2 & 0 & 0 \\ 0 & -t & 0 \\ 0 & 0 & 1 \end{pmatrix}\begin{pmatrix} 1 & -(1+q) & 1 \\ 0 & 1 & -1 \\ 0 & 0 & 1 \end{pmatrix}\begin{pmatrix} q & 0 & 0 \\ 0 & 1 & 0 \\ 0 & 0 & 1 \end{pmatrix} = \begin{pmatrix} t^2q & -t^2(1+q) & t^2 \\ 0 & -t & t \\ 0 & 0 & 1 \end{pmatrix}.$$

$$(13.92)$$

We denote

$$D_{1,2}(q) = \operatorname{diag}(q,1,1) \quad \text{and} \quad D_{2,2}(q) = \operatorname{diag}(1,1,q). \tag{13.93}$$

Similarly, we get

$$\sigma_2 \mapsto \begin{pmatrix} 1 & 0 \\ 1 & 1 \end{pmatrix}\begin{pmatrix} 1 & 0 \\ 0 & -t \end{pmatrix} \overset{\text{Sym}^2}{\mapsto} \begin{pmatrix} 1 & 0 & 0 \\ 1 & 1 & 0 \\ 1 & 2 & 1 \end{pmatrix}\begin{pmatrix} 1 & 0 & 0 \\ 0 & -t & 0 \\ 0 & 0 & t^2 \end{pmatrix} \overset{\text{Quant}}{\mapsto}$$

$$\begin{pmatrix} 1 & 0 & 0 \\ 1 & 1 & 0 \\ 1 & (1+q) & q \end{pmatrix}\begin{pmatrix} 1 & 0 & 0 \\ 0 & 1 & 0 \\ 0 & 0 & q \end{pmatrix}\begin{pmatrix} 1 & 0 & 0 \\ 0 & -t & 0 \\ 0 & 0 & t^2 \end{pmatrix} = \begin{pmatrix} 1 & 0 & 0 \\ 1 & -t & 0 \\ 1 & -t(1+q) & t^2q \end{pmatrix}.$$

$$(13.94)$$

Finally, we have

$$\left[S^2(\sigma_1)_e\right]_q := \begin{pmatrix} t^2q & -t^2(1+q) & t^2 \\ 0 & -t & t \\ 0 & 0 & 1 \end{pmatrix},$$

$$\left[S^2(\sigma_2)e\right]_q := \begin{pmatrix} 1 & 0 & 0 \\ 1 & -t & 0 \\ 1 & -t(1+q) & t^2q \end{pmatrix}. \tag{13.95}$$

To enumerate the elements of the bases e and v by the same set of indices, set

$$w_{ij} = v_{i,j-1}. \tag{13.96}$$

Since the bases e and w (hence v) are connected as follows,

$$\begin{aligned}
e^s_{11} &= w_{11}, & w_{11} &= e^s_{11}, \\
e^s_{12} &= -w_{11} + w_{12} - w_{22}, & w_{12} &= e^s_{11} + e^s_{12} + e^s_{22}, \\
e^s_{22} &= w_{22}, & w_{22} &= e^s_{22},
\end{aligned} \tag{13.97}$$

the *change-of-basis matrix* C_3 in the space $S^2(\mathbb{C}^2)$ is the following:

$$C_3 = \begin{pmatrix} 1 & -1 & 0 \\ 0 & 1 & 0 \\ 0 & -1 & 1 \end{pmatrix}, \quad C_3^{-1} = \begin{pmatrix} 1 & 1 & 0 \\ 0 & 1 & 0 \\ 0 & 1 & 1 \end{pmatrix}. \tag{13.98}$$

We show that

$$C_3\left[S^2(\sigma_r)e\right]_q C_3^{-1} = k^{(t,q)}(\sigma_r), \quad \text{for} \quad r = 1, 2. \tag{13.99}$$

Here, by (13.12), we have $\sigma_{31} = e \in S_3$, which is why $s_{\sigma_{31}} = I$. Indeed, we have

$$\left[S^2(\sigma_1)v\right]_q := C_3 \begin{pmatrix} t^2q & -t^2(1+q) & t^2 \\ 0 & -t & t \\ 0 & 0 & 1 \end{pmatrix} C_3^{-1} = \begin{pmatrix} t^2q & 0 & t(t-1) \\ 0 & 0 & t \\ 0 & 1 & 1-t \end{pmatrix}, \tag{13.100}$$

$$\left[S^2(\sigma_2)v\right]_q := C_3 \begin{pmatrix} 1 & 0 & 0 \\ 1 & -t & 0 \\ 1 & -t(1+q) & t^2q \end{pmatrix} C_3^{-1} = \begin{pmatrix} 0 & t & 0 \\ 1 & 1-t & 0 \\ 0 & qt(t-1) & t^2q \end{pmatrix}. \tag{13.101}$$

This coincides with *the Krammer representation $k_3^{(t,q)}$ for B_3* in notation (13.2):

$$k_3^{(t,q)}(\sigma_1) = \begin{pmatrix} t^2q & 0 & t(t-1) \\ 0 & 0 & t \\ 0 & 1 & 1-t \end{pmatrix}, \quad k_3^{(t,q)}(\sigma_2) = \begin{pmatrix} 0 & t & 0 \\ 1 & 1-t & 0 \\ 0 & qt(t-1) & t^2q \end{pmatrix}.$$

$$(13.102)$$

The quantisation of the symmetric square $S^2(\rho_{n+1}^{(t)})$ of the Burau representations $\rho_{n+1}^{(t)}$ for the group B_{n+1} will be defined as follows, comparing with (13.43):

$$\sigma_1 \to [S^2(\sigma_1)]_q = S^2\big(\exp(sE_{11})\big)\big[S^2\big(\exp(-E_{12})\big)\big]_q D_{1,n}(q),$$

$$(13.103)$$

$$\sigma_n \to [S^2\big(\exp(E_{n-1,n})\big)]_q D_{n,n}(q)S^2\big(\exp(sE_{nn})\big), \tag{13.104}$$

$$\sigma_k \to [S^2\big(\exp(E_{k,k-1})\big)]_q D_{k,n}(q)S^2\big(\exp(sE_{kk})\big)\big[S^2\big(\exp(-E_{kk+1})\big)\big]_q,$$

$$(13.105)$$

where $D_{k,n}(q)$ is defined for $1 \le k \le n$ and $1 \le r \le l \le n$ as follows:

$$D_{k,n}(q)e_{r,l}^s = qe_{r,l}^s, \quad \text{if } k = r = l, \quad \text{and} \quad D_{k,n}(q)e_{r,l}^s = e_{r,l}^s, \quad \text{otherwise.}$$

$$(13.106)$$

The Burau representation $\rho_4^{(t)}$ for B_4 is as follows:

$$\sigma_1 \mapsto \begin{pmatrix} -t & t & 0 \\ 0 & 1 & 0 \\ 0 & 0 & 1 \end{pmatrix}, \quad \sigma_2 \mapsto \begin{pmatrix} 1 & 0 & 0 \\ 1 & -t & t \\ 0 & 0 & 1 \end{pmatrix}, \quad \sigma_3 \mapsto \begin{pmatrix} 1 & 0 & 0 \\ 0 & 1 & 0 \\ 0 & 1 & -t \end{pmatrix}.$$

$$(13.107)$$

The symmetric square of the reduced Burau representation $\rho_4^{(t)}$ for B_4 is as follows, referring to (13.40) and (13.41):

$$S^2(\sigma_1) \mapsto \begin{pmatrix} t^2 & 0 & 0 & 0 & 0 & 0 \\ 0 & -t & 0 & 0 & 0 & 0 \\ 0 & 0 & 1 & 0 & 0 & 0 \\ 0 & 0 & 0 & -t & 0 & 0 \\ 0 & 0 & 0 & 0 & 1 & 0 \\ 0 & 0 & 0 & 0 & 0 & 1 \end{pmatrix} \begin{pmatrix} 1 & -2 & 1 & 0 & 0 & 0 \\ 0 & 1 & -1 & 0 & 0 & 0 \\ 0 & 0 & 1 & 0 & 0 & 0 \\ 0 & 0 & 0 & 1 & -1 & 0 \\ 0 & 0 & 0 & 0 & 1 & 0 \\ 0 & 0 & 0 & 0 & 0 & 1 \end{pmatrix}$$

$$= \begin{pmatrix} t^2 & -2t^2 & t^2 & 0 & 0 & 0 \\ 0 & -t & t & 0 & 0 & 0 \\ 0 & 0 & 1 & 0 & 0 & 0 \\ 0 & 0 & 0 & -t & t & 0 \\ 0 & 0 & 0 & 0 & 1 & 0 \\ 0 & 0 & 0 & 0 & 0 & 1 \end{pmatrix},$$

$$S^2(\sigma_2) = \begin{pmatrix} 1 & 0 & 0 & 0 & 0 & 0 \\ 1 & 1 & 0 & 0 & 0 & 0 \\ 1 & 2 & 1 & 0 & 0 & 0 \\ 0 & 0 & 0 & 1 & 0 & 0 \\ 0 & 0 & 0 & 1 & 1 & 0 \\ 0 & 0 & 0 & 0 & 0 & 1 \end{pmatrix} \begin{pmatrix} 1 & 0 & 0 & 0 & 0 & 0 \\ 0 & -t & 0 & 0 & 0 & 0 \\ 0 & 0 & t^2 & 0 & 0 & 0 \\ 0 & 0 & 0 & 1 & 0 & 0 \\ 0 & 0 & 0 & 0 & -t & 0 \\ 0 & 0 & 0 & 0 & 0 & 1 \end{pmatrix}$$

$$\times \begin{pmatrix} 1 & 0 & 0 & 0 & 0 & 0 \\ 0 & 1 & 0 & -1 & 0 & 0 \\ 0 & 0 & 1 & 0 & -2 & 1 \\ 0 & 0 & 0 & 1 & 0 & 0 \\ 0 & 0 & 0 & 0 & 1 & -1 \\ 0 & 0 & 0 & 0 & 0 & 1 \end{pmatrix} = \begin{pmatrix} 1 & 0 & 0 & 0 & 0 & 0 \\ 1 & -t & 0 & t & 0 & 0 \\ 1 & -2t & t^2 & 2t & -2t^2 & t^2 \\ 0 & 0 & 0 & 1 & 0 & 0 \\ 0 & 0 & 0 & 1 & -t & t \\ 0 & 0 & 0 & 0 & 0 & 1 \end{pmatrix},$$

$$S^2(\sigma_3) \mapsto \begin{pmatrix} 1 & 0 & 0 & 0 & 0 & 0 \\ 0 & 1 & 0 & 0 & 0 & 0 \\ 0 & 0 & 1 & 0 & 0 & 0 \\ 0 & 1 & 0 & 1 & 0 & 0 \\ 0 & 0 & 1 & 0 & 1 & 0 \\ 0 & 0 & 1 & 0 & 2 & 1 \end{pmatrix} \begin{pmatrix} 1 & 0 & 0 & 0 & 0 & 0 \\ 0 & 1 & 0 & 0 & 0 & 0 \\ 0 & 0 & 1 & 0 & 0 & 0 \\ 0 & 0 & 0 & -t & 0 & 0 \\ 0 & 0 & 0 & 0 & -t & 0 \\ 0 & 0 & 0 & 0 & 0 & t^2 \end{pmatrix}$$

$$= \begin{pmatrix} 1 & 0 & 0 & 0 & 0 & 0 \\ 0 & 1 & 0 & 0 & 0 & 0 \\ 0 & 0 & 1 & 0 & 0 & 0 \\ 0 & 1 & 0 & -t & 0 & 0 \\ 0 & 0 & 1 & 0 & -t & 0 \\ 0 & 0 & 1 & 0 & -2t & t^2 \end{pmatrix}.$$

Using formulas (13.103), (13.104) and (13.105), we get

$$S^2(\sigma_1) \overset{\text{Quant}}{\mapsto} \begin{pmatrix} t^2 & 0 & 0 & 0 & 0 & 0 \\ 0 & -t & 0 & 0 & 0 & 0 \\ 0 & 0 & 1 & 0 & 0 & 0 \\ 0 & 0 & 0 & -t & 0 & 0 \\ 0 & 0 & 0 & 0 & 1 & 0 \\ 0 & 0 & 0 & 0 & 0 & 1 \end{pmatrix} \begin{pmatrix} q & -(1+q) & 1 & 0 & 0 & 0 \\ 0 & 1 & -1 & 0 & 0 & 0 \\ 0 & 0 & 1 & 0 & 0 & 0 \\ 0 & 0 & 0 & 1 & 1 & 0 \\ 0 & 0 & 0 & 0 & 1 & 0 \\ 0 & 0 & 0 & 0 & 0 & 1 \end{pmatrix}$$

$$= \begin{pmatrix} t^2 q & -t^2(1+q) & t^2 & 0 & 0 & 0 \\ 0 & -t & t & 0 & 0 & 0 \\ 0 & 0 & 1 & 0 & 0 & 0 \\ 0 & 0 & 0 & -t & t & 0 \\ 0 & 0 & 0 & 0 & 1 & 0 \\ 0 & 0 & 0 & 0 & 0 & 1 \end{pmatrix},$$

$$S^2(\sigma_3) \overset{\text{Quant}}{\longmapsto} \begin{pmatrix} 1 & 0 & 0 & 0 & 0 & 0 \\ 0 & 1 & 0 & 0 & 0 & 0 \\ 0 & 0 & 1 & 0 & 0 & 0 \\ 0 & 1 & 0 & 1 & 0 & 0 \\ 0 & 0 & 1 & 0 & 1 & 0 \\ 0 & 0 & 1 & 0 & (1+q) & q \end{pmatrix} \begin{pmatrix} 1 & 0 & 0 & 0 & 0 & 0 \\ 0 & 1 & 0 & 0 & 0 & 0 \\ 0 & 0 & 1 & 0 & 0 & 0 \\ 0 & 0 & 0 & -t & 0 & 0 \\ 0 & 0 & 0 & 0 & -t & 0 \\ 0 & 0 & 0 & 0 & 0 & t^2 \end{pmatrix}$$

$$= \begin{pmatrix} 1 & 0 & 0 & 0 & 0 & 0 \\ 0 & 1 & 0 & 0 & 0 & 0 \\ 0 & 0 & 1 & 0 & 0 & 0 \\ 0 & 1 & 0 & -t & 0 & 0 \\ 0 & 0 & 1 & 0 & -t & 0 \\ 0 & 0 & 1 & 0 & -t(1+q) & t^2 q \end{pmatrix}.$$

$$S^2(\sigma_2) \overset{\text{Quant}}{\longmapsto} \begin{pmatrix} 1 & 0 & 0 & 0 & 0 & 0 \\ 1 & 1 & 0 & 0 & 0 & 0 \\ 1 & 1+q & 1 & 0 & 0 & 0 \\ 0 & 0 & 0 & 1 & 0 & 0 \\ 0 & 0 & 0 & 1 & 1 & 0 \\ 0 & 0 & 0 & 0 & 0 & 1 \end{pmatrix} \begin{pmatrix} 1 & 0 & 0 & 0 & 0 & 0 \\ 0 & -t & 0 & 0 & 0 & 0 \\ 0 & 0 & qt^2 & 0 & 0 & 0 \\ 0 & 0 & 0 & 1 & 0 & 0 \\ 0 & 0 & 0 & 0 & -t & 0 \\ 0 & 0 & 0 & 0 & 0 & 1 \end{pmatrix}$$

$$\times \begin{pmatrix} 1 & 0 & 0 & 0 & 0 & 0 \\ 0 & 1 & 0 & -1 & 0 & 0 \\ 0 & 0 & 1 & 0 & -(1+q) & 1 \\ 0 & 0 & 0 & 1 & 0 & 0 \\ 0 & 0 & 0 & 0 & 1 & -1 \\ 0 & 0 & 0 & 0 & 0 & 1 \end{pmatrix}$$

$$= \begin{pmatrix} 1 & 0 & 0 & 0 & 0 & 0 \\ 1 & -t & 0 & 0 & 0 & 0 \\ 1 & -t(1+q) & qt^2 & 0 & 0 & 0 \\ 0 & 0 & 0 & 1 & 0 & 0 \\ 0 & 0 & 0 & 1 & -t & 0 \\ 0 & 0 & 0 & 0 & 0 & 1 \end{pmatrix} \begin{pmatrix} 1 & 0 & 0 & 0 & 0 & 0 \\ 0 & 1 & 0 & -1 & 0 & 0 \\ 0 & 0 & 1 & 0 & -(1+q) & 1 \\ 0 & 0 & 0 & 1 & 0 & 0 \\ 0 & 0 & 0 & 0 & 1 & -1 \\ 0 & 0 & 0 & 0 & 0 & 1 \end{pmatrix}$$

$$= \begin{pmatrix} 1 & 0 & 0 & 0 & 0 & 0 \\ 1 & -t & 0 & t & 0 & 0 \\ 1 & -t(1+q) & t^2 q & t(1+q) & -t^2(1+q) & t^2 \\ 0 & 0 & 0 & 1 & 0 & 0 \\ 0 & 0 & 0 & 1 & -t & t \\ 0 & 0 & 0 & 0 & 0 & 1 \end{pmatrix}.$$

Finally, we get in the basis e,

$$\left[S^2(\sigma_1)e\right]_q := \begin{pmatrix} t^2 q & -t^2(1+q) & t^2 & 0 & 0 & 0 \\ 0 & -t & t & 0 & 0 & 0 \\ 0 & 0 & 1 & 0 & 0 & 0 \\ 0 & 0 & 0 & -t & t & 0 \\ 0 & 0 & 0 & 0 & 1 & 0 \\ 0 & 0 & 0 & 0 & 0 & 1 \end{pmatrix},$$

$$\left[S^2(\sigma_3)e\right]_q := \begin{pmatrix} 1 & 0 & 0 & 0 & 0 & 0 \\ 0 & 1 & 0 & 0 & 0 & 0 \\ 0 & 0 & 1 & 0 & 0 & 0 \\ 0 & 1 & 0 & -t & 0 & 0 \\ 0 & 0 & 1 & 0 & -t & 0 \\ 0 & 0 & 1 & 0 & -t(1+q) & t^2 q \end{pmatrix}, \tag{13.108}$$

$$\left[S^2(\sigma_2)_e\right]_q := \begin{pmatrix} 1 & 0 & 0 & 0 & 0 & 0 \\ 1 & -t & 0 & t & 0 & 0 \\ 1 & -t(1+q) & t^2q & t(1+q) & -t^2(1+q) & t^2 \\ 0 & 0 & 0 & 1 & 0 & 0 \\ 0 & 0 & 0 & 1 & -t & t \\ 0 & 0 & 0 & 0 & 0 & 1 \end{pmatrix}.$$

$$(13.109)$$

We rewrite (13.108) and (13.109) in the basis v and compare them with the expressions $k_4^{(t,q)}(\sigma_r)$, for $r = 1, 2, 3$. We maintain the following order to enumerate the elements of the bases e and w in $S^2(\mathbb{C}^3)$:

$$e = (e_{11}^s, e_{12}^s, e_{22}^s, e_{13}^s, e_{23}^s, e_{33}^s), \quad w = (w_{11}, w_{12}, w_{22}, w_{13}, w_{23}, w_{33}).$$

$$(13.110)$$

By Lemma 13.5, the change-of-basis matrices C_4 and C_4^{-1} in the space $S^2(\mathbb{C}^4)$ are the following:

$$C_4 = \begin{pmatrix} 1 & -1 & 0 & 0 & 0 & 0 \\ 0 & 1 & 0 & -1 & 0 & 0 \\ 0 & -1 & 1 & 1 & -1 & 0 \\ 0 & 0 & 0 & 1 & 0 & 0 \\ 0 & 0 & 0 & -1 & 1 & 0 \\ 0 & 0 & 0 & 0 & -1 & 1 \end{pmatrix}, \quad C_4^{-1} = \begin{pmatrix} 1 & 1 & 0 & 1 & 0 & 0 \\ 0 & 1 & 0 & 1 & 0 & 0 \\ 0 & 1 & 1 & 1 & 1 & 0 \\ 0 & 0 & 0 & 1 & 0 & 0 \\ 0 & 0 & 0 & 1 & 1 & 0 \\ 0 & 0 & 0 & 1 & 1 & 1 \end{pmatrix}.$$

$$(13.111)$$

We calculate $\left[S^2(\sigma_r)_v\right]_q = C_4\left[S^2(\sigma_r)_e\right]_q C_4^{-1}$ for $r = 1, 2, 3$. We have

$$\left[S^2(\sigma_1)_v\right]_q = C_4 \begin{pmatrix} t^2q & -t^2(1+q) & t^2 & 0 & 0 & 0 \\ 0 & -t & t & 0 & 0 & 0 \\ 0 & 0 & 1 & 0 & 0 & 0 \\ 0 & 0 & 0 & -t & t & 0 \\ 0 & 0 & 0 & 0 & 1 & 0 \\ 0 & 0 & 0 & 0 & 0 & 1 \end{pmatrix} C_4^{-1}$$

$$= \begin{pmatrix} qt^2 & 0 & t(t-1) & 0 & t(t-1) & 0 \\ 0 & 0 & t & 0 & 0 & 0 \\ 0 & 1 & 1-t & 0 & 0 & 0 \\ 0 & 0 & 0 & 0 & t & 0 \\ 0 & 0 & 0 & 1 & 1-t & 0 \\ 0 & 0 & 0 & 0 & 0 & 1 \end{pmatrix}, \qquad (13.112)$$

$$\left[S^2(\sigma_2)_v\right]_q = C_4 \begin{pmatrix} 1 & 0 & 0 & 0 & 0 & 0 \\ 1 & -t & 0 & t & 0 & 0 \\ 1 & -t(1+q) & t^2q & t(1+q) & -t^2(1+q) & t^2 \\ 0 & 0 & 0 & 1 & 0 & 0 \\ 0 & 0 & 0 & 1 & -t & t \\ 0 & 0 & 0 & 0 & 0 & 1 \end{pmatrix} C_4^{-1}$$

$$= \begin{pmatrix} 0 & t & 0 & 0 & 0 & 0 \\ 1 & 1-t & 0 & 0 & 0 & 0 \\ 0 & qt(t-1) & qt^2 & 0 & 0 & t(t-1) \\ 0 & 0 & 0 & 1 & 0 & 0 \\ 0 & 0 & 0 & 0 & 0 & t \\ 0 & 0 & 0 & 0 & 1 & 1-t \end{pmatrix},$$

$$\left[S^2(\sigma_3)_v\right]_q = C_4 \begin{pmatrix} 1 & 0 & 0 & 0 & 0 & 0 \\ 0 & 1 & 0 & 0 & 0 & 0 \\ 0 & 0 & 1 & 0 & 0 & 0 \\ 0 & 1 & 0 & -t & 0 & 0 \\ 0 & 0 & 1 & 0 & -t & 0 \\ 0 & 0 & 1 & 0 & -t(1+q) & t^2q \end{pmatrix} C_4^{-1}$$

$$
= \begin{pmatrix}
1 & 0 & 0 & 0 & 0 & 0 \\
0 & 0 & 0 & t & 0 & 0 \\
0 & 1 & 0 & 0 & t & 0 \\
0 & 0 & 1 & 1-t & 0 & 0 \\
0 & 0 & 0 & 0 & 1-t & 0 \\
0 & 0 & 0 & qt(t-1) & qt(t-1) & qt^2
\end{pmatrix}.
\tag{13.113}
$$

Applying the operator $s_{\sigma_{41}}$ defined by (13.9), where σ_{41} is defined by (13.19), referring to (13.110), we get $s_{\sigma_{41}}^{-1}\big[S^2(\sigma_r)v\big]_q s_{\sigma_{41}} = k_4^{(t,q)}(\sigma_r)$ for $1 \le r \le 3$. This coincides with *the LK representation* $k_4^{(t,q)}(\sigma_r)$ *for* B_4 in notation (13.2):

$$
k_4^{(t,q)}(\sigma_1) = \begin{pmatrix}
qt^2 & 0 & 0 & t(t-1) & t(t-1) & 0 \\
0 & 0 & 0 & t & 0 & 0 \\
0 & 0 & 0 & 0 & t & 0 \\
0 & 1 & 0 & 1-t & 0 & 0 \\
0 & 0 & 1 & 0 & 1-t & 0 \\
0 & 0 & 0 & 0 & 0 & 1
\end{pmatrix},
$$

$$
k_4^{(t,q)}(\sigma_2) = \begin{pmatrix}
0 & t & 0 & 0 & 0 & 0 \\
1 & 1-t & 0 & 0 & 0 & 0 \\
0 & 0 & 1 & 0 & 0 & 0 \\
0 & qt(t-1) & 0 & qt^2 & 0 & t(t-1) \\
0 & 0 & 0 & 0 & 0 & t \\
0 & 0 & 0 & 0 & 1 & 1-t
\end{pmatrix},
$$

$$
k_4^{(t,q)}(\sigma_3) = \begin{pmatrix}
1 & 0 & 0 & 0 & 0 & 0 \\
0 & 0 & t & 0 & 0 & 0 \\
0 & 1 & 1-t & 0 & 0 & 0 \\
0 & 0 & 0 & 0 & t & 0 \\
0 & 0 & 0 & 1 & 1-t & 0 \\
0 & 0 & qt(t-1) & 0 & qt(t-1) & qt^2
\end{pmatrix}.
$$

Remark 13.12. Kashaev pointed out to the author, during his visit to Geneva in May 2022, that formulas (13.112)–(13.113) should be corrected.

Lemma 13.5. *The matrices C_{n+1}^{-1} and C_{n+1} are defined as follows. The columns of the matrix C_{n+1}^{-1} are defined by the following relations:*

$$w_{ij} = \sum_{i \le k \le r \le j} e_{kr}^s, \quad 1 \le i \le j \le n. \tag{13.114}$$

The columns of the matrix C_{n+1} are defined by the following relations:

$$e_{ij}^s = \begin{cases} (-1)^{i+j} \sum_{i \le k \le r \le j}(-1)^{k+r} w_{kr}, & \text{if } 0 \le j - i \le 1, \\ \sum_{r=i+1}^{j-1}(-w_{ir} + w_{i+1,r}) + w_{ij} - w_{i+1,j}, & \text{if } i = 1, \ 2 \le j - i, \\ -w_{ii} + \sum_{r=i+1}^{j-1}(-w_{ir} + w_{i+1,r}) + w_{ij} - w_{i+1,j}, & \text{if } i > 1, \ 2 \le j - i. \end{cases}$$
$$\tag{13.115}$$

Proof. Indeed, to prove (13.98), (13.111) and (13.119), we note that matrix $E_n = C_n^{-1}$ is block upper-triangular:

$$E_n = \begin{pmatrix} a_n & b_n \\ 0 & c_n \end{pmatrix}, \quad \text{therefore} \quad C_n = E_n^{-1} = \begin{pmatrix} a_n^{-1} & -a_n^{-1} b_n c_n^{-1} \\ 0 & c_n^{-1} \end{pmatrix}.$$
$$\tag{13.116}$$

For $n = 3$, we have

$$E_3 = \begin{pmatrix} 1 & 1 & 0 \\ 0 & 1 & 0 \\ 0 & 1 & 1 \end{pmatrix} = \begin{pmatrix} a_3 & b_3 \\ 0 & c_3 \end{pmatrix}, \quad E_3^{-1} = \begin{pmatrix} a_3^{-1} & -a_3^{-1} b_3 c_3^{-1} \\ 0 & c_3^{-1} \end{pmatrix},$$

where

$$a_3 = (1), \quad b_3 = (1 \ 0), \quad c_3 = \begin{pmatrix} 1 & 0 \\ 1 & 1 \end{pmatrix}.$$

We have $c_3^{-1} = \begin{pmatrix} 1 & 0 \\ -1 & 1 \end{pmatrix}$; therefore, $a_3^{-1}b_3c_3^{-1} = (1)(1\ 0)\begin{pmatrix} 1 & 0 \\ -1 & 1 \end{pmatrix} = (1\ 0)$,

and hence $C_3 = \begin{pmatrix} 1 & -1 & 0 \\ 0 & 1 & 0 \\ 0 & -1 & 1 \end{pmatrix}$. For $n = 4$, we get

$$E_4 = \begin{pmatrix} 1 & 1 & 0 & 1 & 0 & 0 \\ 0 & 1 & 0 & 1 & 0 & 0 \\ 0 & 1 & 1 & 1 & 1 & 0 \\ 0 & 0 & 0 & 1 & 0 & 0 \\ 0 & 0 & 0 & 1 & 1 & 0 \\ 0 & 0 & 0 & 1 & 1 & 1 \end{pmatrix} = \begin{pmatrix} a_4 & b_4 \\ 0 & c_4 \end{pmatrix}, \quad E_4^{-1} = \begin{pmatrix} a_4^{-1} & -a_4^{-1}b_4c_4^{-1} \\ 0 & c_4^{-1} \end{pmatrix},$$

where

$$a_4 = E_3, \quad b_4 = \begin{pmatrix} 1 & 0 & 0 \\ 1 & 0 & 0 \\ 1 & 1 & 0 \end{pmatrix}, \quad c_4 = \begin{pmatrix} 1 & 0 & 0 \\ 1 & 1 & 0 \\ 1 & 1 & 1 \end{pmatrix}.$$

Since $c_4^{-1} = \begin{pmatrix} 1 & 0 & 0 \\ -1 & 1 & 0 \\ 0 & -1 & 1 \end{pmatrix}$, we get

$$a_4^{-1}b_4c_4^{-1} = \begin{pmatrix} 1 & -1 & 0 \\ 0 & 1 & 0 \\ 0 & -1 & 1 \end{pmatrix}\begin{pmatrix} 1 & 0 & 0 \\ 1 & 0 & 0 \\ 1 & 1 & 1 \end{pmatrix}\begin{pmatrix} 1 & 0 & 0 \\ -1 & 1 & 0 \\ 0 & -1 & 1 \end{pmatrix} = \begin{pmatrix} 0 & 0 & 0 \\ 1 & 0 & 0 \\ -1 & 1 & 0 \end{pmatrix},$$

$$C_4 = \begin{pmatrix} 1 & -1 & 0 & 0 & 0 & 0 \\ 0 & 1 & 0 & -1 & 0 & 0 \\ 0 & -1 & 1 & 1 & -1 & 0 \\ 0 & 0 & 0 & 1 & 0 & 0 \\ 0 & 0 & 0 & -1 & 1 & 0 \\ 0 & 0 & 0 & 0 & -1 & 1 \end{pmatrix}.$$

For $n = 5$, we have

$$E_5 = C_5^{-1} = \left(\begin{array}{ccc|ccc|cccc} 1 & 1 & 0 & 1 & 0 & 0 & 1 & 0 & 0 & 0 \\ \hline 0 & 1 & 0 & 1 & 0 & 0 & 1 & 0 & 0 & 0 \\ 0 & 1 & 1 & 1 & 1 & 0 & 1 & 1 & 0 & 0 \\ \hline 0 & 0 & 0 & 1 & 0 & 0 & 1 & 0 & 0 & 0 \\ 0 & 0 & 0 & 1 & 1 & 0 & 1 & 1 & 0 & 0 \\ 0 & 0 & 0 & 1 & 1 & 1 & 1 & 1 & 1 & 0 \\ \hline 0 & 0 & 0 & 0 & 0 & 0 & 1 & 0 & 0 & 0 \\ 0 & 0 & 0 & 0 & 0 & 0 & 1 & 1 & 0 & 0 \\ 0 & 0 & 0 & 0 & 0 & 0 & 1 & 1 & 1 & 0 \\ 0 & 0 & 0 & 0 & 0 & 0 & 1 & 1 & 1 & 1 \end{array}\right) = \begin{pmatrix} a_5 & b_5 \\ 0 & c_5 \end{pmatrix},$$

$$E_5^{-1} = \begin{pmatrix} a_5^{-1} & -a_5^{-1}b_5c_5^{-1} \\ 0 & c_5^{-1} \end{pmatrix},$$

where

$$a_5 = E_4, \quad b_5 = \begin{pmatrix} 1 & 0 & 0 & 0 \\ 1 & 0 & 0 & 0 \\ 1 & 1 & 0 & 0 \\ 1 & 0 & 0 & 0 \\ 1 & 1 & 0 & 0 \\ 1 & 1 & 1 & 0 \end{pmatrix}, \quad c_5 = \begin{pmatrix} 1 & 0 & 0 & 0 \\ 1 & 1 & 0 & 0 \\ 1 & 1 & 1 & 0 \\ 1 & 1 & 1 & 1 \end{pmatrix}.$$

Since $c_5^{-1} = \begin{pmatrix} 1 & 0 & 0 & 0 \\ -1 & 1 & 0 & 0 \\ 0 & -1 & 1 & 0 \\ 0 & 0 & -1 & 1 \end{pmatrix}$, we get

$$a_5^{-1}b_5c_5^{-1} = \begin{pmatrix} 1 & -1 & 0 & 0 & 0 & 0 \\ 0 & 1 & 0 & -1 & 0 & 0 \\ 0 & -1 & 1 & 1 & -1 & 0 \\ 0 & 0 & 0 & 1 & 0 & 0 \\ 0 & 0 & 0 & -1 & 1 & 0 \\ 0 & 0 & 0 & 0 & -1 & 1 \end{pmatrix} \begin{pmatrix} 1 & 0 & 0 & 0 \\ 1 & 0 & 0 & 0 \\ 1 & 1 & 0 & 0 \\ 1 & 0 & 0 & 0 \\ 1 & 1 & 0 & 0 \\ 1 & 1 & 1 & 0 \end{pmatrix} \begin{pmatrix} 1 & 0 & 0 & 0 \\ -1 & 1 & 0 & 0 \\ 0 & -1 & 1 & 0 \\ 0 & 0 & -1 & 1 \end{pmatrix}$$

$$= \begin{pmatrix} 0 & 0 & 0 & 0 \\ 1 & 0 & 0 & 0 \\ -1 & 1 & 0 & 0 \\ 1 & 0 & 0 & 0 \\ -1 & 1 & 0 & 0 \\ 0 & -1 & 1 & 0 \end{pmatrix},$$

and

$$C_5 = \left(\begin{array}{ccc|ccc|cccc} 1| & -1 & 0| & 0 & 0 & 0| & 0 & 0 & 0 & 0 \\ 0| & 1 & 0| & -1 & 0 & 0| & -1 & 0 & 0 & 0 \\ 0| & -1 & 1| & 1 & -1 & 0| & 1 & -1 & 0 & 0 \\ \hline 0| & 0 & 0| & 1 & 0 & 0| & -1 & 0 & 0 & 0 \\ 0| & 0 & 0| & -1 & 1 & 0| & 1 & -1 & 0 & 0 \\ 0| & 0 & 0| & 0 & -1 & 1| & 0 & 1 & -1 & 0 \\ \hline 0| & 0 & 0| & 0 & 0 & 0| & 1 & 0 & 0 & 0 \\ 0| & 0 & 0| & 0 & 0 & 0| & -1 & 1 & 0 & 0 \\ 0| & 0 & 0| & 0 & 0 & 0| & 0 & -1 & 1 & 0 \\ 0| & 0 & 0| & 0 & 0 & 0| & 0 & 0 & -1 & 1 \end{array}\right).$$

$\square$

Remark 13.13. The matrices $E_n = C_n^{-1}$ and $E_n^{-1} = C_n$ have the following form:

$$E_n = \begin{pmatrix} e_{11} & e_{12} & e_{13} & \cdots & e_{1n-1} \\ 0 & e_{22} & e_{23} & \cdots & e_{2n-1} \\ 0 & 0 & e_{33} & \cdots & e_{3n-1} \\ & & & \vdots & \\ 0 & 0 & 0 & \cdots & e_{n-1,n-1} \end{pmatrix},$$

$$E_n^{-1} = \begin{pmatrix} e_{11}^{-1} & e_{12}^{-1} & e_{13}^{-1} & \cdots & e_{1n-1}^{-1} \\ 0 & e_{22}^{-1} & e_{23}^{-1} & \cdots & e_{2n-1}^{-1} \\ 0 & 0 & e_{33}^{-1} & \cdots & e_{3n-1}^{-1} \\ & & & \vdots & \\ 0 & 0 & 0 & \cdots & e_{n-1,n-1}^{-1} \end{pmatrix}, \qquad (13.117)$$

where $e_{kr} \in \mathrm{Mat}(k \times r, \mathbb{C})$ for $1 \le k \le r \le n - 1$ are as follows:

$$e_{11} = (1), \quad e_{22} = \begin{pmatrix} 1 & 0 \\ 1 & 1 \end{pmatrix}, \quad e_{33} = \begin{pmatrix} 1 & 0 & 0 \\ 1 & 1 & 0 \\ 1 & 1 & 1 \end{pmatrix},$$

$$e_{kk} = (I - e_k)^{-1}, \quad e_k = \sum_{r=1}^{k-1} E_{r+1,r},$$

$e_{kr} = (e_{kk}, 0_{k,r-k})$, where $0_{k,r} = 0$ in $\mathrm{Mat}(k \times r, \mathbb{C})$. For E_n^{-1}, we have

$$e_{11}^{-1} = (1), \quad e_{22}^{-1} = \begin{pmatrix} 1 & 0 \\ -1 & 1 \end{pmatrix}, \quad e_{33}^{-1} = \begin{pmatrix} 1 & 0 & 0 \\ -1 & 1 & 0 \\ 0 & -1 & 1 \end{pmatrix}, \quad e_{kk} = (I - e_k),$$

and $e_{1r}^{-1} = 0$ for $2 \le r \le n - 1$, $e_{kr}^{-1} := (-e_{kk}^{-1}, 0_{k,r-k})$ for $2 \le k < r \le n - 1$. The notation e_{kr}^{-1} for $k < r$ does not mean the inverse matrix to e_{kr}!

The Burau representation $\rho_5^{(t)}$ for B_5 is as follows:

$$
\sigma_1 \mapsto \begin{pmatrix} -t & t & 0 & 0 \\ 0 & 1 & 0 & 0 \\ 0 & 0 & 1 & 0 \\ 0 & 0 & 0 & 1 \end{pmatrix}, \quad
\sigma_2 \mapsto \begin{pmatrix} 1 & 0 & 0 & 0 \\ 1 & -t & t & 0 \\ 0 & 0 & 1 & 0 \\ 0 & 0 & 0 & 1 \end{pmatrix},
$$

$$
\sigma_3 \mapsto \begin{pmatrix} 1 & 0 & 0 & 0 \\ 0 & 1 & 0 & 0 \\ 0 & 1 & -t & t \\ 0 & 0 & 0 & 1 \end{pmatrix}, \quad
\sigma_4 \mapsto \begin{pmatrix} 1 & 0 & 0 & 0 \\ 0 & 1 & 0 & 0 \\ 0 & 0 & 1 & 0 \\ 0 & 0 & 1 & -t \end{pmatrix}.
\tag{13.118}
$$

By Lemma 13.5, the change-of-basis matrices C_5 and C_5^{-1} in the space $S^2(\mathbb{C}^5)$ are the following:

$$
C_5 = \begin{pmatrix}
1 & -1 & 0 & 0 & 0 & 0 & 0 & 0 & 0 & 0 \\
0 & 1 & 0 & -1 & 0 & 0 & -1 & 0 & 0 & 0 \\
0 & -1 & 1 & 1 & -1 & 0 & 1 & -1 & 0 & 0 \\
0 & 0 & 0 & 1 & 0 & 0 & -1 & 0 & 0 & 0 \\
0 & 0 & 0 & -1 & 1 & 0 & 1 & -1 & 0 & 0 \\
0 & 0 & 0 & 0 & -1 & 1 & 0 & 1 & -1 & 0 \\
0 & 0 & 0 & 0 & 0 & 0 & 1 & 0 & 0 & 0 \\
0 & 0 & 0 & 0 & 0 & 0 & -1 & 1 & 0 & 0 \\
0 & 0 & 0 & 0 & 0 & 0 & 0 & -1 & 1 & 0 \\
0 & 0 & 0 & 0 & 0 & 0 & 0 & 0 & -1 & 1
\end{pmatrix},
$$

$$C_5^{-1} = \begin{pmatrix} 1 & 1 & 0 & 1 & 0 & 0 & 1 & 0 & 0 & 0 \\ 0 & 1 & 0 & 1 & 0 & 0 & 1 & 0 & 0 & 0 \\ 0 & 1 & 1 & 1 & 1 & 0 & 1 & 1 & 0 & 0 \\ 0 & 0 & 0 & 1 & 0 & 0 & 1 & 0 & 0 & 0 \\ 0 & 0 & 0 & 1 & 1 & 0 & 1 & 1 & 0 & 0 \\ 0 & 0 & 0 & 1 & 1 & 1 & 1 & 1 & 1 & 0 \\ 0 & 0 & 0 & 0 & 0 & 0 & 1 & 0 & 0 & 0 \\ 0 & 0 & 0 & 0 & 0 & 0 & 1 & 1 & 0 & 0 \\ 0 & 0 & 0 & 0 & 0 & 0 & 1 & 1 & 1 & 0 \\ 0 & 0 & 0 & 0 & 0 & 0 & 1 & 1 & 1 & 1 \end{pmatrix}. \tag{13.119}$$

Using (13.2), we get

$$k_5^{(t,q)}(\sigma_3) = \begin{pmatrix} 1 & 0 & 0 & 0 & 0 & 0 & 0 & 0 & 0 & 0 \\ 0 & 0 & t & 0 & 0 & 0 & 0 & 0 & 0 & 0 \\ 0 & 1 & 1-t & 0 & 0 & 0 & 0 & 0 & 0 & 0 \\ 0 & 0 & 0 & 1 & 0 & 0 & 0 & 0 & 0 & 0 \\ 0 & 0 & 0 & 0 & 0 & t & 0 & 0 & 0 & 0 \\ 0 & 0 & 0 & 0 & 1 & 1-t & 0 & 0 & 0 & 0 \\ 0 & 0 & 0 & 0 & 0 & 0 & 1 & 0 & 0 & 0 \\ 0 & 0 & qt(t-1) & 0 & 0 & qt(t-1) & 0 & qt^2 & 0 & t(t-1) \\ 0 & 0 & 0 & 0 & 0 & 0 & 0 & 0 & 0 & t \\ 0 & 0 & 0 & 0 & 0 & 0 & 0 & 0 & 1 & 1-t \end{pmatrix}.$$

Indeed, since

$$
\sigma_3 \mapsto \begin{pmatrix} 1 & 0 & 0 & 0 \\ 0 & 1 & 0 & 0 \\ 0 & 1 & -t & t \\ 0 & 0 & 0 & 1 \end{pmatrix}
$$

$$
= \begin{pmatrix} 1 & 0 & 0 & 0 \\ 0 & 1 & 0 & 0 \\ 0 & 1 & 1 & 0 \\ 0 & 0 & 0 & 1 \end{pmatrix} \begin{pmatrix} 1 & 0 & 0 & 0 \\ 0 & 1 & 0 & 0 \\ 0 & 0 & -t & 0 \\ 0 & 0 & 0 & 1 \end{pmatrix} \begin{pmatrix} 1 & 0 & 0 & 0 \\ 0 & 1 & 0 & 0 \\ 0 & 0 & 1 & -1 \\ 0 & 0 & 0 & 1 \end{pmatrix}
$$

$$
= \exp(E_{32}) \exp(sE_{33}) \exp(-E_{34}),
$$

and

$$
S^2(e^{E_{32}}) = \begin{pmatrix}
1 & 0 & 0 & 0 & 0 & 0 & 0 & 0 & 0 & 0 \\
0 & 1 & 0 & 0 & 0 & 0 & 0 & 0 & 0 & 0 \\
0 & 1 & 1 & 0 & 0 & 0 & 0 & 0 & 0 & 0 \\
0 & 1 & 0 & 1 & 0 & 0 & 0 & 0 & 0 & 0 \\
0 & 0 & 1 & 0 & 1 & 0 & 0 & 0 & 0 & 0 \\
0 & 0 & 1 & 0 & 2 & 1 & 0 & 0 & 0 & 0 \\
0 & 0 & 0 & 0 & 0 & 0 & 1 & 0 & 0 & 0 \\
0 & 0 & 0 & 0 & 0 & 0 & 0 & 1 & 0 & 0 \\
0 & 0 & 0 & 0 & 0 & 0 & 0 & 1 & 1 & 0 \\
0 & 0 & 0 & 0 & 0 & 0 & 0 & 0 & 0 & 1
\end{pmatrix},
$$

$$
S^2(e^{-E_{34}}) = \begin{pmatrix}
1 & 0 & 0 & 0 & 0 & 0 & 0 & 0 & 0 & 0 \\
0 & 1 & 0 & 0 & 0 & 0 & 0 & 0 & 0 & 0 \\
0 & 0 & 1 & 0 & 0 & 0 & 0 & 0 & 0 & 0 \\
0 & 0 & 0 & 1 & 0 & 0 & -1 & 0 & 0 & 0 \\
0 & 0 & 0 & 0 & 1 & 0 & 0 & -1 & 0 & 0 \\
0 & 0 & 0 & 0 & 0 & 1 & 0 & 0 & -2 & 1 \\
0 & 0 & 0 & 0 & 0 & 0 & 1 & 0 & 0 & 0 \\
0 & 0 & 0 & 0 & 0 & 0 & 0 & 1 & 0 & 0 \\
0 & 0 & 0 & 0 & 0 & 0 & 0 & 0 & 1 & -1 \\
0 & 0 & 0 & 0 & 0 & 0 & 0 & 0 & 0 & 1
\end{pmatrix},
$$

we get

$$S^2(\sigma_3) \overset{\text{Quant}}{\mapsto} \begin{pmatrix} 1 & 0 & 0 & 0 & 0 & 0 & 0 & 0 & 0 & 0 \\ 0 & 1 & 0 & 0 & 0 & 0 & 0 & 0 & 0 & 0 \\ 0 & 1 & 1 & 0 & 0 & 0 & 0 & 0 & 0 & 0 \\ 0 & 1 & 0 & 1 & 0 & 0 & 0 & 0 & 0 & 0 \\ 0 & 0 & 1 & 0 & 1 & 0 & 0 & 0 & 0 & 0 \\ 0 & 0 & 1 & 0 & (1+q) & 1 & 0 & 0 & 0 & 0 \\ 0 & 0 & 0 & 0 & 0 & 0 & 1 & 0 & 0 & 0 \\ 0 & 0 & 0 & 0 & 0 & 0 & 0 & 1 & 0 & 0 \\ 0 & 0 & 0 & 0 & 0 & 0 & 0 & 1 & 1 & 0 \\ 0 & 0 & 0 & 0 & 0 & 0 & 0 & 0 & 0 & 1 \end{pmatrix}$$

$$\times \begin{pmatrix} 1 & 0 & 0 & 0 & 0 & 0 & 0 & 0 & 0 & 0 \\ 0 & 1 & 0 & 0 & 0 & 0 & 0 & 0 & 0 & 0 \\ 0 & 0 & 1 & 0 & 0 & 0 & 0 & 0 & 0 & 0 \\ 0 & 0 & 0 & -t & 0 & 0 & 1 & 0 & 0 & 0 \\ 0 & 0 & 0 & 0 & -t & 0 & 0 & 0 & 0 & 0 \\ 0 & 0 & 0 & 0 & 0 & qt^2 & 0 & 0 & 0 & 1 \\ 0 & 0 & 0 & 0 & 0 & 0 & 1 & 0 & 0 & 0 \\ 0 & 0 & 0 & 0 & 0 & 0 & 0 & 1 & 0 & 0 \\ 0 & 0 & 0 & 0 & 0 & 0 & 0 & 0 & -t & 0 \\ 0 & 0 & 0 & 0 & 0 & 0 & 0 & 0 & 0 & 1 \end{pmatrix}$$

$$\times \begin{pmatrix} 1 & 0 & 0 & 0 & 0 & 0 & 0 & 0 & 0 & 0 \\ 0 & 1 & 0 & 0 & 0 & 0 & 0 & 0 & 0 & 0 \\ 0 & 0 & 1 & 0 & 0 & 0 & 0 & 0 & 0 & 0 \\ 0 & 0 & 0 & 1 & 0 & 0 & -1 & 0 & 0 & 0 \\ 0 & 0 & 0 & 0 & 1 & 0 & 0 & -1 & 0 & 0 \\ 0 & 0 & 0 & 0 & 0 & 1 & 0 & 0 & -(1+q) & 1 \\ 0 & 0 & 0 & 0 & 0 & 0 & 1 & 0 & 0 & 0 \\ 0 & 0 & 0 & 0 & 0 & 0 & 0 & 1 & 0 & 0 \\ 0 & 0 & 0 & 0 & 0 & 0 & 0 & 0 & 1 & -1 \\ 0 & 0 & 0 & 0 & 0 & 0 & 0 & 0 & 0 & 1 \end{pmatrix}$$

$$= \begin{pmatrix}
1 & 0 & 0 & 0 & 0 & 0 & 0 & 0 & 0 & 0 \\
0 & 1 & 0 & 0 & 0 & 0 & 0 & 0 & 0 & 0 \\
0 & 1 & 1 & 0 & 0 & 0 & 0 & 0 & 0 & 0 \\
0 & 0 & 0 & -t & 0 & 0 & 0 & 0 & 0 & 0 \\
0 & 0 & 1 & 0 & -t & 0 & 0 & 0 & 0 & 0 \\
0 & 0 & 1 & 0 & -t(1+q) & qt^2 & 0 & 0 & 0 & 0 \\
0 & 0 & 0 & 0 & 0 & 0 & 1 & 0 & 0 & 0 \\
0 & 0 & 0 & 0 & 0 & 0 & 0 & 1 & 0 & 0 \\
0 & 0 & 0 & 0 & 0 & 0 & 0 & 1 & -t & 0 \\
0 & 0 & 0 & 0 & 0 & 0 & 0 & 0 & 0 & 1
\end{pmatrix}$$

$$\times \begin{pmatrix}
1 & 0 & 0 & 0 & 0 & 0 & 0 & 0 & 0 & 0 \\
0 & 1 & 0 & 0 & 0 & 0 & 0 & 0 & 0 & 0 \\
0 & 0 & 1 & 0 & 0 & 0 & 0 & 0 & 0 & 0 \\
0 & 0 & 0 & 1 & 0 & 0 & -1 & 0 & 0 & 0 \\
0 & 0 & 0 & 0 & 1 & 0 & 0 & -1 & 0 & 0 \\
0 & 0 & 0 & 0 & 0 & 1 & 0 & 0 & -(1+q) & 1 \\
0 & 0 & 0 & 0 & 0 & 0 & 1 & 0 & 0 & 0 \\
0 & 0 & 0 & 0 & 0 & 0 & 0 & 1 & 0 & 0 \\
0 & 0 & 0 & 0 & 0 & 0 & 0 & 0 & 1 & -1 \\
0 & 0 & 0 & 0 & 0 & 0 & 0 & 0 & 0 & 1
\end{pmatrix}$$

$$= \begin{pmatrix}
1 & 0 & 0 & 0 & 0 & 0 & 0 & 0 & 0 & 0 \\
0 & 1 & 0 & 0 & 0 & 0 & 0 & 0 & 0 & 0 \\
0 & t & 1 & 0 & 0 & 0 & 0 & 0 & 0 & 0 \\
0 & 0 & 0 & -t & 0 & 0 & t & 0 & 0 & 0 \\
0 & 0 & 1 & 0 & -t & 0 & 0 & t & 0 & 0 \\
0 & 0 & 1 & 0 & -t(1+q) & qt^2 & 0 & t(1+q) & -t^2(1+q) & t^2 \\
0 & 0 & 0 & 0 & 0 & 0 & 1 & 0 & 0 & 0 \\
0 & 0 & 0 & 0 & 0 & 0 & 0 & 1 & 0 & 0 \\
0 & 0 & 0 & 0 & 0 & 0 & 0 & 1 & -t & t \\
0 & 0 & 0 & 0 & 0 & 0 & 0 & 0 & 0 & 1
\end{pmatrix}.$$

Applying the operator $s_{\sigma_{51}}$, defined by (13.9), where σ_{51} is defined by (13.19), we get $s_{\sigma_{51}}^{-1} C_5 \left[S^2(\sigma_3)_v \right]_q C_5^{-1} s_{\sigma_{51}} = k_5^{(t,q)}(\sigma_3)$. We use the order in the basis $S^2(\mathbb{C}^4)$ as in (13.110). Indeed,

$$s_{\sigma_{51}}^{-1} C_5 \left[S^2(\rho_5^{(t)}(\sigma_3)_e \right]_q C_5^{-1} s_{\sigma_{51}}$$

$$= s_{\sigma_{51}} \begin{pmatrix}
1 & -1 & 0 & 0 & 0 & 0 & 0 & 0 & 0 & 0 \\
0 & 1 & 0 & -1 & 0 & 0 & -1 & 0 & 0 & 0 \\
0 & -1 & 1 & 1 & -1 & 0 & 1 & -1 & 0 & 0 \\
0 & 0 & 0 & 1 & 0 & 0 & -1 & 0 & 0 & 0 \\
0 & 0 & 0 & -1 & 1 & 0 & 1 & -1 & 0 & 0 \\
0 & 0 & 0 & 0 & -1 & 1 & 0 & 1 & -1 & 0 \\
0 & 0 & 0 & 0 & 0 & 0 & 1 & 0 & 0 & 0 \\
0 & 0 & 0 & 0 & 0 & 0 & -1 & 1 & 0 & 0 \\
0 & 0 & 0 & 0 & 0 & 0 & 0 & -1 & 1 & 0 \\
0 & 0 & 0 & 0 & 0 & 0 & 0 & 0 & -1 & 1
\end{pmatrix}$$

$$\times \begin{pmatrix}
1 & 0 & 0 & 0 & 0 & 0 & 0 & 0 & 0 & 0 \\
0 & 1 & 0 & 0 & 0 & 0 & 0 & 0 & 0 & 0 \\
0 & t & 1 & 0 & 0 & 0 & 0 & 0 & 0 & 0 \\
0 & 0 & 0 & -t & 0 & 0 & t & 0 & 0 & 0 \\
0 & 0 & 1 & 0 & -t & 0 & 0 & t & 0 & 0 \\
0 & 0 & 1 & 0 & -t(1+q) & qt^2 & 0 & t(1+q) & -t^2(1+q) & t^2 \\
0 & 0 & 0 & 0 & 0 & 0 & 1 & 0 & 0 & 0 \\
0 & 0 & 0 & 0 & 0 & 0 & 0 & 1 & 0 & 0 \\
0 & 0 & 0 & 0 & 0 & 0 & 0 & 1 & -t & t \\
0 & 0 & 0 & 0 & 0 & 0 & 0 & 0 & 0 & 1
\end{pmatrix}$$

$$\times \begin{pmatrix} 1 & 1 & 0 & 1 & 0 & 0 & 1 & 0 & 0 & 0 \\ 0 & 1 & 0 & 1 & 0 & 0 & 1 & 0 & 0 & 0 \\ 0 & 1 & 1 & 1 & 1 & 0 & 1 & 1 & 0 & 0 \\ 0 & 0 & 0 & 1 & 0 & 0 & 1 & 0 & 0 & 0 \\ 0 & 0 & 0 & 1 & 1 & 0 & 1 & 1 & 0 & 0 \\ 0 & 0 & 0 & 1 & 1 & 1 & 1 & 1 & 1 & 0 \\ 0 & 0 & 0 & 0 & 0 & 0 & 1 & 0 & 0 & 0 \\ 0 & 0 & 0 & 0 & 0 & 0 & 1 & 1 & 0 & 0 \\ 0 & 0 & 0 & 0 & 0 & 0 & 1 & 1 & 1 & 0 \\ 0 & 0 & 0 & 0 & 0 & 0 & 1 & 1 & 1 & 1 \end{pmatrix} s_{\sigma_{51}}^{-1}$$

$$= \begin{pmatrix} 1 & 0 & 0 & 0 & 0 & 0 & 0 & 0 & 0 & 0 \\ 0 & 0 & t & 0 & 0 & 0 & 0 & 0 & 0 & 0 \\ 0 & 1 & 1-t & 0 & 0 & 0 & 0 & 0 & 0 & 0 \\ 0 & 0 & 0 & 1 & 0 & 0 & 0 & 0 & 0 & 0 \\ 0 & 0 & 0 & 0 & 0 & t & 0 & 0 & 0 & 0 \\ 0 & 0 & 0 & 0 & 1 & 1-t & 0 & 0 & 0 & 0 \\ 0 & 0 & 0 & 0 & 0 & 0 & 1 & 0 & 0 & 0 \\ 0 & 0 & qt(t-1) & 0 & 0 & t(t-1) & 0 & qt^2 & 0 & t(t-1) \\ 0 & 0 & 0 & 0 & 0 & 0 & 0 & 0 & 0 & t \\ 0 & 0 & 0 & 0 & 0 & 0 & 0 & 0 & 1 & 1-t \end{pmatrix}$$

$$\overset{(13.2)}{=} k_5^{(t,q)}(\sigma_3).$$

The formulas (13.7) can be obtained similarly.

Remark 13.14. We can prove (13.87) for general B_n. The reason is the following. By Remark 12.1, we can construct the reduced Burau representation for B_{n+1} from the reduced Burau representation for B_n using formula (12.13). In particular, we can construct the reduced Burau representation for B_5, starting from the reduced Burau representation for B_4 using formulas (12.14).

By Lemma 13.1, we can reconstruct the LK representation $k_{n+1}^{(t,q)}$ for B_{n+1} from the LK representation $k_n^{(t,q)}$ for B_n, defined by (13.2). That is why it is sufficient to prove Theorem 13.1 only for B_3 and B_4.

Chapter 14

How to Construct New Representations of B_n

14.1 Quantisation of the mth Symmetric Power of the Burau Representation

We begin with the well-known representation (in fact, reduced Burau representation for $t = -1$) $T_{32} := \rho_3^{(-1)} : B_3 \to \mathrm{SL}(2, \mathbb{Z})$, defined by (13.88):

$$\sigma_1 \mapsto \begin{pmatrix} 1 & -1 \\ 0 & 1 \end{pmatrix}, \quad \sigma_2 \mapsto \begin{pmatrix} 1 & 0 \\ 1 & 1 \end{pmatrix}, \quad \text{or} \quad \sigma_1 \mapsto \begin{pmatrix} 1 & 1 \\ 0 & 1 \end{pmatrix}, \quad \sigma_2 \mapsto \begin{pmatrix} 1 & 0 \\ -1 & 1 \end{pmatrix}. \tag{14.1}$$

This representation is simply the exponential of the fundamental (or natural) representations π_2 of the Lie algebra $\mathfrak{sl}_2$; see (13.38) and (13.39).

Remark 14.1. We have *several possibilities* to generalise the representation T_{32} of the group B_3. *First*, we can generalise T_{32} for an arbitrary B_n (denoted by T_{n2}); *second*, we can generalise T_{32} of the same group B_3 but for an arbitrary dimension m (denoted by T_{3m}), and *finally*, we can generalise the representations T_{n2} of B_n for a higher dimension m, (denoted by T'_{nm}) and then extend the representation T_{3m} of B_3 for other groups B_n (denoted by T''_{nm}). Finally, we have two families of representations:

$$T'_{nm}, T''_{nm} : B_n \to \mathrm{GL}(m, \mathbb{C}).$$

217

To be more precise, the *first step* gives us all the Burau representations, $\rho_{n+1}^{(t)} : B_{n+1} \to \mathrm{GL}_n(\mathbb{Z}[t,t^{-1}])$, due to Theorem 13.2, if we start with $\rho_3^{(t)}$ and take for π to be the natural representation π_n of $\mathfrak{sl}_n$.

In the *second step*, we obtain the Humphries representation of B_3 in all dimensions $m+1$ as the mth symmetric power of the Burau representation ρ_3^{-1} due to Theorem 13.2. Remarkably, that *quantisation* and *diagonal deformation* (governed by the matrix $\Lambda_n = \mathrm{diag}(\lambda_r)_{r=0}^n$, see Remark 13.5) of the Humphries representations, referring to (13.59), *include all irreducible representations* of B_3 in dimension ≤ 5 due to the results of Tuba (2001) and Tuba and Wenzl (2001) and our equivalent description of their representations (Kosyak, 2008); see Theorem 13.4. Our approach using the q-Pascal triangle allows us to freely extend the representation T_{32} of B_3 for an arbitrary dimension; see Theorem 13.3.

In the *third step*, we obtain first the Lawrence–Krammer representation as a quantization of the symmetric square of the Burau representation, as give by Theorem 13.1. Due to formulas (13.79) and (13.80) and Theorem 13.1, the Lawrence–Krammer representations of B_3 describe all irreducible representations, up to the scalar factor, in dimension 3.

Problem 14.1. It would be nice to obtain an explicit formula for the quantisation of the symmetric cube and higher symmetric powers of the Burau representation, as given in the following table:

$V = \mathbb{C}^n$	B_3	B_n	Type of Representation
2	$T_{3,2}^{\Lambda_1,1,\pi_2}$	$\rho_n^{(t)}$	Reduced Burau representation
3	$T_{3,3}^{\Lambda_2,q,\pi_2}$	$k_n^{(t,q)}$	Lawrence–Krammer representation
4	$T_{3,4}^{\Lambda_3,q,\pi_2}$	$[S^3(\rho_n^{(t)})]_q$	Symmetric cube
5	$T_{3,5}^{\Lambda_4,q,\pi_2}$	$[S^4(\rho_n^{(t)})]_q$	4th symmetric power
$m+1$	$T_{3,m+1}^{\Lambda_m,q,\pi_2}$	$[S^m(\rho_n^{(t)})]_q$	mth symmetric power

14.2 Representations of the Lie Algebra $\mathfrak{sl}_{n+1}$

We recall, following (Serre, 1964, Part I, Ch. VII, §1; Part III, Ch. VII, §6), that all irreducible finite-dimensional representations of a *simple* Lie algebra $\mathfrak{sl}_{n+1}$ are the *highest weight representations* and are contained in tensor powers of the *natural representation* π_{n+1} (Weyl, 1939).

The *Cartan subalgebra* $\mathfrak{h}$ of the Lie algebra $\mathfrak{sl}_{n+1}$ consists of all matrices of the form $H = \mathrm{diag}(\lambda_1, \ldots, \lambda_{n+1})$, with $\sum_k \lambda_k = 0$. The *roots* of a Lie algebra $\mathfrak{sl}_{n+1}$ are linear functionals $\alpha_{i,j} \in \mathfrak{h}^*$, $i \neq j$, of the form

$$\alpha_{i,j}(H) = \lambda_i - \lambda_j, \quad \text{where} \quad H = \mathrm{diag}(\lambda_1, \ldots, \lambda_{n+1}).$$

The *fundamental weights* ω_k are defined by

$$\omega_k(H) = \lambda_1 + \cdots + \lambda_k, \quad \text{where} \quad H = \mathrm{diag}(\lambda_1, \ldots, \lambda_{n+1}).$$

The fundamental weight ω_1 is that of the *standard representation* π_{n+1} of the Lie algebra $\mathfrak{sl}_{n+1}$ in $E = \mathbb{C}^{n+1}$. Moreover, the fundamental weight ω_k is the highest weight of the representation $\wedge^k(\pi_{n+1})$ of $\mathfrak{sl}_{n+1}$ in the space $\wedge^k E$ (see Serre, 1964, 1.7.4.6), where $\wedge^k(A)$ is the kth *exterior power* of A.

In a particular case, all irreducible representations of $\mathfrak{sl}_2$ are $\mathrm{Sym}^n(\pi_2)$ in $\mathrm{Sym}^n(\mathbb{C}^2)$, $n \in \mathbb{N}$. For $\mathfrak{sl}_3$, all irreducible representations are contained in the tensor product of π_3 and $\wedge^2(\pi_3)$, corresponding to the fundamental weights ω_1 and ω_2. All irreducible representations of $\mathfrak{sl}_{n+1}$ are contained in the tensor products of $\wedge^k(\pi_{n+1})$, $1 \leq k \leq n$.

14.3 New Representations of B_n

Let $T_{nm} : \mathfrak{sl}_n \to \mathrm{End}(\mathbb{C}^m)$ be any finite-dimensional representation of a Lie algebra $\mathfrak{sl}_n$. Then, $\exp(T_{nm}) : \mathrm{SL}(n, \mathbb{C}) \to \mathrm{GL}(m, \mathbb{C})$ is a representation of $\mathrm{SL}(n, \mathbb{C})$. We can define the representations of B_{n+1} as follows:

$$\sigma_k \to S^r\big(\exp(T_{nm}) \circ \rho_n^{(t)}(\sigma_k)\big), \quad \sigma_k \to \wedge^l\big(\exp(T_{nm}) \circ \rho_n^{(t)}(\sigma_k)\big),$$

$$(14.2)$$

for $r \geq 2$ and $2 \leq l \leq m-1$.

Consider the exterior square $\wedge^2(\rho_{2,4}^{(t)})$ of the Burau representation $\rho_{2,4}^{(t)}$ of the group B_4, defined by (12.4) and (12.5), in the basis $e_{ij}^{\wedge}$, $1 \leq i < j \leq 3$ of the space $\wedge^2(\mathbb{C}^3)$:

$$e_{12}^{\wedge} = e_1 \otimes e_2 - e_2 \otimes e_1, \quad e_{13}^{\wedge} = e_1 \otimes e_3 - e_3 \otimes e_1, \quad e_{23}^{\wedge} = e_2 \otimes e_3 - e_3 \otimes e_2.$$

Lemma 14.1. *The representation $\wedge^2(\rho_{2,4}^{(t)})$ of the group B_4 is defined by*

$$\sigma_1 \to \begin{pmatrix} -t & 0 & 0 \\ 0 & -t & t \\ 0 & 0 & 1 \end{pmatrix}, \quad \sigma_2 \to \begin{pmatrix} -t & t & 0 \\ 0 & 1 & 0 \\ 0 & 1 & -t \end{pmatrix}, \quad \sigma_3 \to \begin{pmatrix} 1 & 0 & 0 \\ 1 & -t & 0 \\ 0 & 0 & -t \end{pmatrix}.$$

$$(14.3)$$

We have

$$\wedge^2 \rho_{2,4}^{(t)} \not\sim \rho_4^{(t)}, \quad \text{for} \quad t \neq -1, \quad \text{but} \quad \wedge^2 \rho_{2,4}^{(t)} = -t J_2 \rho_{1,4}^{(-t^{-1})} J_2^{-1},$$

$$(14.4)$$

where $\rho_{1,4}^{(t)}$ is defined by (12.2) and (12.3), and $J_2 = \sum_{k=1}^{3} E_{k,4-k} = \begin{pmatrix} 0\,0\,1 \\ 0\,1\,0 \\ 1\,0\,0 \end{pmatrix}$.

Proof. The first part of (14.4) follows from a comparison of two spectra: $\mathrm{Sp}\,\rho_{2,4}^{(t)}(\sigma_1) = \{-t, 1, 1\}$ and $\mathrm{Sp}(\wedge^2 \rho_{2,4}^{(t)}(\sigma_1)) = \{-t, 1, -t\} = -t\{1, -t^{-1}, 1\}$. The second part follows from (12.2) and (12.3). $\qquad\square$

It would be nice to quantise and study the representation of B_n, defined by (14.2):

$$\sigma_k \to \big[S^r\big(\exp(T_{nm}) \circ \rho_n^{(t)}(\sigma_k)\big)\big]_q, \quad \sigma_k \to \big[\wedge^l\big(\exp(T_{nm}) \circ \rho_n^{(t)}(\sigma_k)\big)\big]_q.$$

$$(14.5)$$

Bibliography

Albeverio, S. and Kosyak, A. (2007). q-Pascal's triangle and irreducible representations of the braid group B_3 in arbitrary dimension, arXiv:math.QA(RT)/0803.2778v2.

Andrews, G. E. (1976). *The Theory of Partitions*, Encyclopedia of Mathematics and Its Applications, Addison–Wesley Publishing Company, Cambridge, Massachusetts.

Artin, E. (1926). Theorie des Zöpfe, *Abh. Math. Sem. Hamburg. Univ.* **4**, pp. 47–72.

Bigelow, S. (1999). The Burau representation of the braid group B_n is not faithful for n = 5, *Geom. Topol.* **3**, pp. 397–404.

Bigelow, S. (2001). Braid groups are linear, *J. Amer. Math. Soc.* **14**(2), pp. 471–486.

Bigelow, S. (2002a). The Lawrence–Krammer representation, arXiv:math.GT/0204057v1.

Bigelow, S. (2002b). Does the Jones polynomial detect the unknot? *J. Knot Theory Ramifications* **11**(4), pp. 493–505.

Birman, J. S. (1974). *Braids, Links, and Mapping Class Groups*, Ann. Math. Studies. **82**, Princeton University Press, Princeton, NJ, USA.

Birman, J. S. and Brendel, T. E. (2005). *Braids: A Survey, in: Handbook of Knot Theory*, W. Menasco and T. Thistlethwaite (Eds.), Elsevier, The Netherlands.

Burau, W. (1936). Über Zopfgruppen und gleichsinnig verdrillte Verkettungen. *Abh. Math. Sem. Univ. Hamburg.* **11**, pp. 179–186.

Formanek, E., Woo, W., Sysoeva, I., Vazirani, M. (2003). The irreducible complex representations of n string of degree $\leq n$, *J. Algebra Appl.* **2**, pp. 317–333.

Humphries, S. P. (2000). Some linear representations of braid groups, *J. Knot Theory Ramifications.* **9**(3), pp. 341–366.

Jones, V. F. R. (1987). Hecke algebra representations of braid groups and link polynomials, *Ann. Math.* **126**, pp. 335–388.

Kassel, C. (1995). *Quantum Groups*, Springer-Verlag, New York, NY, USA.

Kassel, C. and Turaev, V. (2008). Braid groups, *Graduate Texts in Mathematics*, Vol. **247**, Springer, New York, (340 p. + xii).

Kosyak, A. V. (2008). Representations of the braid group B_n and the highest weight modules of $U(\mathfrak{sl}_{n-1})$ and $U_q(\mathfrak{sl}_{n-1})$, arXiv:math.QA(RT)0803.2785v2.

Kosyak, A. V. (2025b). The Lawrence-Krammer representation is a quantization of the symmetric square of the Burau representation, *Linear Algebra Appl.* **727**, pp. 203–233, https://doi.org/10.1016/j.laa.2025.08.009.

Krammer, D. (2000). The braid group B4 is linear, *Invent. Math.* **142**(3), pp. 451–486.

Krammer, D. (2002). Braid groups are linear, *Ann. Math.* **155**(2), pp. 131–156.

Larsen, M. J. and Rowell, E. C. (2008). An algebra-level version of a link-polynomial identity of Lickorish, *Math. Proc. Cambridge Philos. Soc.* **144**(3), pp. 623–638.

Lawrence, R. (1990). Homological representations of the Hecke algebra, *Commun. Math. Phys.* **135**, pp. 141–191.

Lickorish, W. B. R. (1989). Some link-polynomial relations, *Math. Proc. Camb. Phil. Soc.* **105**(1), pp. 103–107.

Long, P. and Paton, M. (1993). The Burau representation of the braid group B_n is not faithful for $n \geq 6$, *Topology* **32**, pp. 439–447.

Magnus, W. and Peluso, A. (1969). On a theorem of V.I. Arnold, *Comm. Pure Appl. Math.* **22**, pp. 683–692.

Moody, J. (1991). The Burau representation of the braid group B_n is not faithful for large n, *Bull. Math. Soc.* **25**, pp. 379–384.

Serre, J. -P. (1964). *Lie Algebras and Lie Groups*, 1964 Lectures Given at Harvard University, Part of Lecture Notes in Mathematics, Vol. 1500.

Smeltzer, R. (2003). *Linear Representations of Braid Groups*. A thesis for the degree of the master of sciences, McMaster University, July 2003. 65 p.

Stoimenov, A. (2008). The density of Lawrence–Krammer and non-conjugate braid representation of links, arXiv:math.GR/0809.0033 v2.

Tuba, I. (2001). Low-dimensional unitary representations of B_3, *Proc. Amer. Math. Soc.* **129**, pp. 2597–2606.

Tuba, I. and Wenzl, H. (2001). Representations of the braid group B_3 and of $SL(2, \mathbb{Z})$, *Pacific J. Math.* **197**(2) pp. 491–510.

Wenzl, H. (1990). Quantum groups and subfactors of type B, C, and D, *Comm. Math. Phys.* **133**, pp. 383–432.

Westbury, B. (1995). On the character varieties of the modular group, Preprint, Nottingham, 1995.

Weyl, H. (1939). *The Classical Groups, Their Invariants and Representations*, (First Edition, 1939, Second Edition with supplements, 1949) Princeton University Press, Princeton, NJ, USA.

Zinno, M. (2001). On Krammer's representation of the braid group, *Math. Ann.* **321**, pp. 197–211.

Index